Actuarial Models for Disability Insurance

Actuarial Models for Disability Insurance

S. Haberman
Professor, Department of Actuarial Science,
City University, London

E. Pitacco
Professor, Department of Actuarial Mathematics,
University of Trieste, Italy

CHAPMAN & HALL/CRC

Boca Raton   London   New York   Washington, D.C.

Library of Congress Cataloging-in-Publication Data

Catalog record is available from the Library of Congress.

Contents

Preface

The aims of this book are to describe the mathematics of multiple state models and to illustrate how this powerful, mathematical framework, based on Markov and semi-Markov stochastic processes, can be used to develop a general, unified and rigorous approach for describing and analysing disability and related insurance benefits. We focus on disability and related benefits (rather than conventional life insurance) and, in particular, consider permanent health insurance (as described in the UK), critical illness insurance and long-term care insurance. These types of insurance cover are becoming increasingly important at a time of demographic ageing in many developed countries and as the demand for health-related insurance products is increasing from a public that is becoming better educated and more prosperous. The pressures on existing welfare systems are growing, leading governments in many countries to seek new ways to fund care for the disabled and elderly, often in partnership with the insurance industry.

The use of Markov chains in life contingencies and their extensions has been proposed by several authors, in both the time-continuous case and the time-discrete case; for example, see Amsler (1968, 1988), Haberman (1983, 1984), Hoem (1969, 1988), Jones (1993), Wilkie (1988). The reader interested in applications of semi-Markov models to actuarial (and demographic) problems should consult Hoem (1972); the first application of semi-Markov processes to disability benefits appears in Janssen (1966).

As far as disability benefits (and in particular PHI benefits) are concerned, the mathematics of Markov and semi-Markov chains provides both a powerful modelling tool and a unifying point of view, from which several calculation techniques and conventional procedures can be seen in a new light (see Haberman, 1988; Waters, 1984, 1989; CMIR 12, 1991).

Though an explicit and systematic use of the mathematics of multiple state models dates back to the end of the 1960s, it must be stressed that the basic mathematics of what we now call a Markov chain model were developed during the eighteenth century (see Seal, 1977); seminal contributions by D. Bernoulli and P.S. de Laplace demonstrate this for the continuous-time case. Moreover, the well-known paper of Hamza (1900) provides the actuarial literature with the first systematic approach to disability benefits, in both the continuous and the discrete case, and Du Pasquier (1912, 1913) provides the first presentation and solution of the underlying differential equations for the transition probabilities.

In this book, we present a brief review of the history of the application of multiple state models to disability and related insurances. Chapter 1 presents the basic probabilistic structure in terms of a continuous-time approach, with a general introduction to the use of the multiple state model for insurances of the person, including a consideration of a range of simple actuarial examples, in terms of actuarial values, net premiums and reserves. Chapter 2 takes a discrete time approach and also considers the concepts of emerging costs and profit testing. Disability insurance is considered in Chapter 3 in some detail, with practical calculation procedures as used widely in a number of countries briefly illustrated and reinterpreted in the framework of multiple state modelling. This approach enables useful comparisons to be drawn. Chapter 4 deals with the calculation of transition probabilities from observations and, in particular, with the graduation (or smoothing) of the transition intensities for the multiple state model. Chapter 5 considers the application of multiple state models to critical illness insurance and Chapter 6 to long-term care insurance. Chapter 7 considers the use of multiple state models for measuring the impact of HIV and AIDS on disability insurance. The indexing of benefits and premiums in insurance covers that include disability insurance is discussed in Chapter 8.

In writing this book, we have attempted to maintain an international perspective, recognizing the dynamic nature of this subject and that the development of new coverages has not occurred in one single location. Examples of approaches from a number of markets are described – in particular, in Chapters 3, 5 and 6. However, there is a limit to the amount of practical detail that can be presented in a book such as this and hence we have avoided providing a detailed analysis of policy conditions and also of profit and expense loadings.

The technical level of the book is intended to match the background and needs of final year undergraduate students, graduate teachers and professionals working in the disability and health insurance fields.

We believe that the unifying approach provided by multiple state models is very important from an educational viewpoint and we are pleased to note the extent to which these models now feature in degree courses in Actuarial Science and in the curricula of actuarial professional bodies around the world.

We would like to acknowledge the invaluable assistance of Annamaria Olivieri in the preparation and analysis of the numerical examples that have been used as well as in reading the text and making comments. Of course, the authors are responsible for any remaining errors and omissions.

Introduction: a history of multiple state models and the actuarial contribution to disability insurance

MULTIPLE STATE MODELS

Multiple state and multiple decrement models provide a powerful tool for application in many areas of actuarial science, particularly in the actuarial assessment of sickness insurance and disability income benefits, and as such feature prominently in this book.

The early history of these models has been described by Seal (1977) and Daw (1979) in some detail and our purpose here is merely to outline the key historical developments in terms of the theory and its practical application to insurance problems.

The models can be traced back to Bernoulli (1766), who applied the methods of differential calculus to a particular problem related to smallpox morbidity and mortality and then solved the resulting differential equations under specified constraints. The problem is the following.

> Given two states A and B such that individuals in state A have mutually exclusive probabilities, possibly dependent on the time spent in state A, of leaving that state because of (i) death, or (ii) passage to state B, what is the probability of an individual passing to state B and dying there within a given period? (Seal, 1977)

Bernoulli's state A consisted of individuals who had never had smallpox, while state B comprised those who had contracted smallpox and would either die from it, almost immediately, or survive and no longer be suffering from that disease. In solving this problem, Bernoulli started with Edmund Halley's (Breslau) life table and effectively produced the first double decrement life table together with one of the related single decrement tables, as well as considering the efficacy of inoculation and deriving a mathematical model of the behaviour of smallpox in a community. The approach was based on setting up and solving a series of differential equations. (This work was the forerunner of substantial developments in the mathematical theory of infectious diseases: see Bailey (1975) for further discussion.)

During the next 50 years, there were a number of contributions from other authors on the subject of smallpox and inoculation, including Jean d'Alembert and Jean Trembley.

Lambert (1772: volume III) explained how numerical data could be used to study Bernoulli's problem and laid the practical foundations for the double decrement model and life table (Daw, 1980). He obtained an approximate formula for the rate of mortality when smallpox is excluded, thereby setting down a practical connection between the double decrement model and the underlying single decrement models. Daw (1979) argues that, with Lambert's contribution, 'the practical and theoretical foundations of double decrement tables had been laid down'.

Despite this progress, by the early 1800s there were two outstanding problems, namely (i) deriving accurate practical formulae for application to numerical data, linking the discrete and continuous cases; and (ii) obtaining exact results in a convenient form (d'Alembert had derived an exact result in terms of an integral that was difficult to evaluate).

These problems were attacked successfully and independently by Cournot (1843: Chapter XIII) and by Makeham (1867). They were the first to set down the fundamental relations of multiple decrement models: in modern notation

$$\mu_x^k = (a\mu)_x^k \quad \text{for } k = 1, \ldots, m$$

$$(a\mu)_x = \sum_{k=1}^{m} \mu_x^k$$

from which it follows that

$$_n(ap)_x = \prod_{k=1}^{m} {}_np_x^k.$$

Makeham (1867) also contains an analysis of the 'partial' forces of mortality for different causes of death, suggesting a reinterpretation of his well-known formula for the aggregate force of mortality

$$\mu_x = A + Bc^x = \left(\sum_{i=1}^{n} A_i\right) + \left(\sum_{j=1}^{m} B_j\right)c^x$$

to represent the separate contributions from $m + n$ causes of death.

Makeham went on to use this connection between the forces of decrement to interpret the prior development of the theory; he demonstrated that the earlier results of Bernoulli and d'Alembert satisfied this additive law for the forces of decrement and this multiplicative law for the probabilities (or corresponding l_x functions). He also identified an error made by d'Alembert (Makeham, 1875).

In an internal report in 1875 (which was not placed in the public domain) on the invalidity and widows' pension scheme for railway officials, Karup described the properties and use of single decrement probabilities and forces of decrement in the context of an illness–death model (with no recoveries permitted), i.e. the 'independent or pure' probabilities of mortality and disablement. He also discussed the estimation of the independent probabilities from observed data. This approach was regarded as controversial by many German actuaries and led to a furious debate in the national literature: the arguments are reviewed by Du Pasquier (1913): see below. Karup (1893) provides a definitive review of this work and describes clearly and in full detail the valuation of liabilities and of the contributions for widows' funds, dealing also with pensions payable on invalidity and pensions payable to orphans (using the 'reversionary method'). Sprague (1879) is an important contribution as far as the UK literature is concerned, demonstrating how to estimate directly the forces of decrement and the (independent) single decrement probabilities. Sprague's arguments are intuitive, based on an assumption of a uniform distribution of decrements over each year of age, but are nonetheless effective. Sprague's paper is a standard which has laid the foundations for subsequent practical applications in the English literature, which tended to follow Sprague's lead rather than build on the earlier, more fundamentals work of Makeham.

Hamza (1900) represents an important development by providing a systematic approach to disability benefits in both the continuous and discrete cases. In any scientific endeavour, communication is facilitated by the availability of a concise and agreed notation. In this regard, Hamza's paper is significant, setting down a notation which has been widely adopted in the following decades and which forms the basis for the notation we have utilized in this monograph.

Du Pasquier (1912, 1913) took a dramatic step forward by providing a rigorous, mathematical discussion of the invalidity or sickness process with the introduction of a three-state illness–death model in which recoveries were permitted (to be discussed in detail in Chapter 3). He derived the full differential equations for the transition probabilities and showed that these lead to a second-order differential equation of Riccati type which he then solved for the case of constant forces of transition. The three sections of these papers dealt respectively with a general presentation of the differential equations, the special case where recoveries are not permitted and then the more general case allowing for recoveries. Du Pasquier's work is very significant, presenting an early application of

Markov chains, and laying the foundations for modern actuarial applications to disability insurance, long-term care insurance and critical illness, *inter alia* (Pitacco, 1995a).

The work of Karup, du Pasquier and others in mainland Europe meant that the actuarial theory associated with sickness and invalidity/disability insurance developed there more rapidly than in the UK, where simple methods based on proportions sick (as advocated by Watson's 1903 book – see below) remained in use through the nineteenth century and a good part of the twentieth century. Chapters on the three-state illness–death model can thus be found in the textbooks of Berger (1935) and Zwinggi (1958).

Despite the interest and importance of these problems to actuaries and the consistent contributions made to the actuarial literature since the mid-nineteenth century, these contributions have essentially been rediscovered and renamed (as the theory of competing risks) by Neyman (1950) and by Fix and Neyman (1951), and other statistical workers, who developed the formulae and results *ab initio* in the context of Markov processes, with little reference to their actuarial predecessors.

APPLICATIONS TO INSURANCE

The actuarial contribution to health and sickness insurance is closely linked with the evolution of friendly societies in the UK and corresponding institutions in other countries. The friendly society is a mutual association which gives financial assistance to its members in times of sickness and old age and meets the costs of burial. Its operations are based on insurance principles, the benefits being paid from a fund accumulated from the members' regular contributions.

The origins of friendly societies in the UK may be traced to the craft guilds of the thirteenth and fourteenth centuries. But the most significant factor in their later development was the Industrial Revolution, which was accompanied by widespread poverty and insecurity. At this time, a rapid population increase (attributable to a decline in mortality rates) provided a source of cheap and easily exploitable labour. In the absence of state provision through social insurance schemes and of occupational pension or sick-pay schemes, friendly societies, established through the initiative of working men, were the only form of financial security available. The development of friendly societies in the UK in the nineteenth century was paralleled elsewhere, for example in the USA and France, and was predated by the miners' mutual help societies of Germany and Austria which were set up in the eighteenth century.

In the early days, the financial arrangement of these societies was rudimentary. The fact that the average cost per member of payments for the contingencies covered tends to increase with age was generally ignored.

The contributions were often fixed on the levy principle (i.e. the cost of current claims was met by a uniform charge on all active members, regardless of age), leading to classic instances of adverse selection and financial ruin (Cummins *et al.*, 1982).

The actuarial profession began to be concerned with friendly society finance during the nineteenth century. As their techniques developed, actuaries advised on the rates of contribution and on the accumulation of funds to meet the future liabilities being promised.

The first attempt to produce age-related sickness rates was made by Price (1792) (the fifth edition of his classic book) at the request of a House of Commons Committee. It is not clear whether these related rates are hypothetical or based on the collection and analysis of experience data. The rates were used to produce tables of contributions for given levels of benefit in respect of incapacity for work.

The first investigation into the observed sickness experience of an institution is the report of 1824 from the Highland Society of Scotland. At this time, UK friendly societies were in difficulties, through a failure to understand their financial status; in some, capital funds had increased regularly for a number of years and benefit payments were then increased without there being a full recognition of the prospective implications arising from benefit levels that might already have been too high. The Highland Society's analysis was crude but did include the average number of weeks sick in decennial age groups, interpolated to give rates for each year of age as well as an occupational analysis of the experience. The rates emerging were much lower than those subsequently experienced – one deficiency noted by many writers was the improper treatment of the many withdrawals.

Finlaison (1829) presents a classic investigation of annuities which contains sickness rates based on six years' data (derived from the experience of the London Society up to age 60 and the Highland Society thereafter). Finlaison calculated the present value of a standard set of friendly society benefits at different ages and the equivalent age-dependent contribution rates, noting the implications of charging the same contribution rate for all. This work represents the first English investigation into sickness rates.

Hubbard (1852) contains the first investigation into the sickness experience of friendly societies in France with an analysis by sex and occupation. For example, we find the first published analysis of data on length of hospital stays for each spell of sickness for different types of occupation. Hubbard's work was used subsequently by French friendly societies for the calculation of contribution rates.

In 1855, K.F. Heym published work on the organization of friendly societies, with special reference to Leipzig, in which he advocated the use of premiums which were dependent on age at entry. He also published age-related sickness rates and probabilities of being disabled, based on data collected from local friendly societies, and sets of annual and single premiums for standard patterns of sickness benefit (Lazarus, 1860). Heym is regarded by some commentators as the 'creator of invalidity insurance science' (Seal, 1977).

In 1854, Prussia legislated for the compulsory membership of sick relief societies; this legislation was copied by the other northern states of Germany. In 1871, the owners of railways, mines and factories were made liable by law for injuries (or deaths) caused to the employed by accidents where the sufferers were not themselves culpable. In this context, a significant work is Wiegand (1859), which deals with the reorganization of the invalidity insurance scheme for the German railways. Wiegand based his numerical calculations on the probabilities and rates presented earlier by Heym. He also published an analysis of the mortality and invalidity experience of railway officials and used insurance company data for the costing of invalidity benefits (with adjustments to allow for the select nature of railway workers). In an earlier paragraph, we have mentioned the report of Karup (1893) which also provides a clear presentation of the methodology for valuing invalidity benefits.

Watson (1903) provides an account of investigations into the sickness and mortality experience for 1893–97 of the Manchester Unity Friendly Society which has become one of the standard works of sickness insurance in the UK. Its methodology has dominated UK actuarial practice for almost nine decades. Although methods have now moved on from Watson's approach, it is still possible to appreciate his clear presentation and analysis of data cross-classified by several factors. Earlier investigations had revealed that occupation was an important variable (see above), but Watson's consideration of the combined effect of occupation and region was new.

The friendly society movement has declined in importance during the twentieth century; however, there is now a resurgence of interest in many countries in long-term disability income insurance (or permanent health insurance), long-term care insurance for the elderly, critical illness protection and other types of related cover offered by life insurance companies. It is with these important modern applications in mind that this textbook has been written.

1

Multiple state models for life and other contingencies: the time-continuous approach

1.1 STATES AND TRANSITIONS

The evolution of an insured risk can be viewed as a sequence of events which determine the cash flows of premiums and benefits. When insurances of the person are concerned, examples of such events include disablement, recovery, death, marriage, birth of a child, onset of a particular illness, etc.

We assume that the evolution of a risk can be described in terms of the presence of the risk itself, at every point of time, in a certain **state** belonging to a specified set of states, or **state space**. Furthermore, we assume that the aforementioned events correspond to **transitions** from one state to another state.

The graph of Fig. 1.1 illustrates a set of four states, numbered 1 to 4 (**nodes** of the graph), and a set of possible 'direct' transitions between states, denoted by pairs such as $(1,2)$, $(2,1)$, $(1,3)$, etc. (**arcs**). 'Indirect' transitions can be represented by sequences of arcs: for example, a transition from 2 to 3 can be represented by the sequence $(2,1)$, $(1,3)$.

For instance, let us suppose that state 1 is the state at policy issue. Possible paths of the insured risk are as follows:

$$1 \text{ (until the policy term)};$$

$$1 - 2 - 4;$$

$$1 - 2 - 1 - 3 - 4.$$

From a merely intuitive point of view, it appears that:

- states 1 and 2 are **transient states**: it is possible to leave and to re-enter these states;

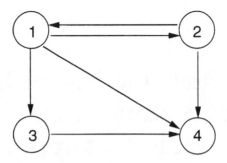

Fig. 1.1 Set of states and set of transitions.

- state 3 is a **strictly transient state**: it is not possible to enter this state once it has been left;
- state 4 is an **absorbing state**: it is not possible to leave this state once it has been entered.

A rigorous definition of these types of states will be given in terms of transition probabilities.

Formally, we denote by $\mathscr{S}$ the state space. We assume that $\mathscr{S}$ is a finite set. Then, denoting the states by integral numbers, we have:

$$\mathscr{S} = \{1, 2, \ldots, N\}. \tag{1.1}$$

The set of direct transitions is denoted by $\mathscr{T}$. In general, $\mathscr{T}$ is a subset of the set of pairs (i, j):

$$\mathscr{T} \subseteq \{(i, j) \mid i \neq j; i, j \in \mathscr{S}\}. \tag{1.2}$$

If state 1 is the initial state at time 0, it is assumed that all the states $j \in \mathscr{S}$ can be reached from state 1 by direct or indirect transitions.

The pair $(\mathscr{S}, \mathscr{T})$ is called a **multiple state model**. Note that a multiple state model $(\mathscr{S}, \mathscr{T})$ simply describes the 'uncertainty', i.e. the 'possibilities' pertaining to an insured risk, as far as its evolution is concerned. A probabilistic structure will be introduced later (see section 1.4), in order to express a numerical evaluation of the uncertainty.

Let us suppose that we are at policy issue, i.e. at time 0. The time unit is one year. Let $S(t)$ denote the random state occupied by the risk at time t, $t \geq 0$. Of course, $S(0)$ is a given state; we can assume, for example, $S(0) = 1$. $\{S(t); t \geq 0\}$ is a **time-continuous stochastic process**, with values in the finite set $\mathscr{S}$. The variable t is often called **seniority**. It represents the duration of the policy; when a single life is concerned, whose age at policy issue is x, $x + t$ represents the attained age. Any possible realization $\{s(t)\}$ of the process $\{S(t)\}$ is called a **sample path**; thus, $s(t)$ is a function of the non-negative variable t, with values in $\mathscr{S}$.

1.2 BENEFITS AND PREMIUMS

As the next step in describing the insurance contract, benefits and premiums must be introduced. First, let us consider Fig. 1.2. The items in Fig. 1.2 are as follows:

$s(t)$ a sample path of the stochastic process $\{S(t)\}$;

$p_1(t)$ a continuous premium, at an instantaneous rate $p_1(t)$, paid by the insured while the risk is in state 1;

$b_2(t)$ a continuous annuity, at an instantaneous rate $b_2(t)$, paid by the insurer while the risk is in state 2;

$c_{13}(t_3)$ a lump sum paid by the insurer at time t_3 because a transition from state 1 to state 3 occurs;

$c_{34}(t_5)$ a lump sum paid by the insurer at time t_5 because a transition from state 3 to state 4 occurs;

$d_3(t_4)$ a lump sum paid by the insurer at fixed time t_4 because the risk is in state 3.

Thus, from the point of view of the insurer, $p_1(t)$ determines an inflow, whilst the bs, the cs and the ds correspond to outflows, i.e. benefits.

In general, the following types of premiums and benefits will be considered:

- a continuous annuity benefit at a rate $b_j(t)$ at time t if $S(t) = j$; $b_j(t)\,dt$ is the benefit amount paid out in the infinitesimal interval $[t, t + dt)$;
- a lump sum, $c_{ij}(t)$ at time t if a transition occurs from state i to state j at that time;
- a lump sum $d_j(t)$ ('pure endowment') at some fixed time t if $S(t) = j$;

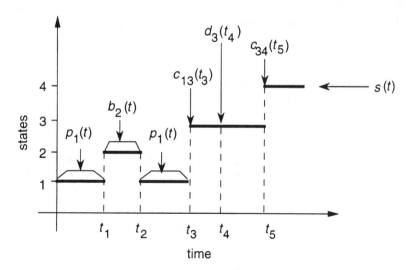

Fig. 1.2 A sample path of $\{S(t)\}$. Benefits and premiums.

- a continuous premium at a rate $p_i(t)$ at time t if $S(t) = i$; $p_i(t)\,dt$ is the premium amount paid in $[t, t + dt)$.

Moreover, discrete-time premiums of the following type can be usefully considered:

- a premium $\pi_i(t)$ at some fixed time t if $S(t) = i$; for instance, as we assume $S(0) = 1$, $\pi_1(0)$ can represent the (initial) single premium (in this case, all other premium functions are identically equal to zero).

Remark 1

For the sake of simplicity, some redundancy has been allowed in defining the premium and benefit functions. A more concise (and more general) notation is as follows (see Wolthuis, 1994). Let $\Pi_i(t)$ and $B_j(t)$ respectively denote the cumulative premium functions and the cumulative annuity benefit functions. Assume that these functions are non-negative and non-decreasing and that (i) premiums are paid by the insured for an amount $\Pi_i(u) - \Pi_i(t)$ while in state i in the interval $[t, u)$; (ii) annuity benefits due by the insurer in state j over the same interval are $B_j(u) - B_j(t)$. Possible jumps in the functions $B_j(t)$ denote pure endowments. When continuous benefits are concerned, it is usual to write $dB_j(t) = b_j(t)\,dt$, where $b_j(t)$ denotes the rate of benefit at time t. Analogous considerations hold for the premium functions $\Pi_i(t)$.

Remark 2

The premium functions defined above can be considered as relating to net premiums or to office premiums; in the latter case, the expense loading is implicitly considered. In a similar way, the benefit functions can implicitly include administrative expenses. Nevertheless, in many applications it is necessary to treat separately net premiums and expense loadings, and, on the other side, benefits and expenses. To do this, further functions should be considered. However, unless stated otherwise, we shall restrict our attention to the study of net premiums and benefits.

1.3 EXAMPLES

In this section we present several examples which illustrate states, transitions and benefits pertaining to insurances of the person.

Example 1

Consider a **temporary assurance**, with a constant sum assured c and continuous premium at a constant rate p; let n denote the term of the policy. The graph describing states and possible transitions is depicted in Fig. 1.3. Then we have

$$c_{12}(t) = c \quad (0 < t \le n)$$

$$p_1(t) = \begin{cases} p & \text{if } 0 \le t < n \\ 0 & \text{if } t \ge n \end{cases}$$

Fig. 1.3 A two-state model.

It is understood that the functions which do not appear in the definition of benefits and premiums (for example, $p_2(t)$) must be considered identically equal to zero. In the case of a single premium, on the other hand, we have $p_1(t) = 0$ and $\pi_1(0) = \pi$.

Example 2

Consider an **endowment assurance,** with c as sum assured in the case of death and in the case of survival to maturity as well. The graph is still given by Fig. 1.3 and we have (for the case of a continuous premium):

$$c_{12}(t) = c \quad (0 < t \leq n)$$
$$d_1(n) = c$$
$$p_1(t) = \begin{cases} p & \text{if } 0 \leq t < n \\ 0 & \text{if } t \geq n \end{cases}.$$

Example 3

Consider a **deferred annuity.** Premiums are assumed to be paid continuously at a rate p over $[0, m)$ when the contract stays in state 1 (i.e. the insured is alive). The benefit is a continuous annuity at a rate b after m when the contract stays in state 1, i.e. until the death of the insured. Thus, benefit and premium functions are as follows:

$$b_1(t) = \begin{cases} 0 & \text{if } 0 \leq t < m \\ b & \text{if } t \geq m \end{cases}$$
$$p_1(t) = \begin{cases} p & \text{if } 0 \leq t < m \\ 0 & \text{if } t \geq m \end{cases}.$$

Example 4

As the next step in building up more complex models, consider a **temporary assurance with a rider benefit in the case of accidental death.** In this case, we have to distinguish between death due to an accident and death due to other causes. The graphical representation is given by Fig. 1.4.

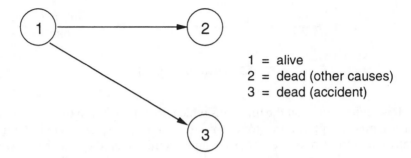

Fig. 1.4 A three-state model with two causes of 'decrement'.

The benefit and premium functions are specified as follows:

$$c_{12}(t) = c \quad (0 < t \leq n)$$

$$c_{13}(t) = c' \quad (0 < t \leq n)$$

$$p_1(t) = \begin{cases} p & \text{if } 0 \leq t < n \\ 0 & \text{if } t \geq n \end{cases},$$

where $c' > c$, and $c' - c$ is the amount of the supplementary benefit in the case of accidental death.

Example 5

Consider now an n-year insurance contract just providing a **lump sum benefit in the case of permanent and total disability**. Also in this case a three-state model is required: states 'active', 'disabled' and 'dead' must be considered. It is important to stress that, since permanent and total disability only is involved, the label 'active' concerns any insured who is alive and non-permanently (or non-totally) disabled. The model is presented in Fig. 1.5. Let c denote the sum assured. Premiums are assumed to be paid continuously at a rate p over $[0, n)$ when the contract stays in state 1 (i.e. when the insured is active).

The benefit and premium functions are as follows:

$$c_{12}(t) = c \quad (0 < t \leq n)$$

$$p_1(t) = \begin{cases} p & \text{if } 0 \leq t < n \\ 0 & \text{if } t \geq n \end{cases}.$$

The above-described model is very simple but rather unrealistic. It is more realistic to assume that the lump sum benefit will be paid out after a qualification period, which is required by the insurer in order to ascertain the permanent character of the disability; the length of the

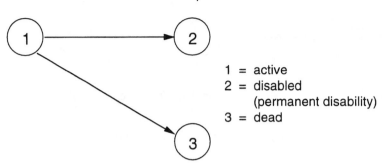

1 = active
2 = disabled
(permanent disability)
3 = dead

Fig. 1.5 A further three-state model with two causes of 'decrement'.

qualification period would be chosen in such a way that recovery would be practically impossible after that period. A more realistic model will be presented in Chapter 3.

Example 6

Examples 4 and 5 can be generalized to include more than two causes of 'decrement'. This leads to the widely discussed **multiple decrement model**. For example, we may be interested in the benefits provided to members within an occupational pension scheme should they retire, die or leave the scheme. A possible model is presented in Fig. 1.6. Note that in this simplified model it is assumed that there is no impact if death occurs after retirement or withdrawal.

Example 7

A more complicated structure than Example 1 can be used to represent mortality due to a certain disease. In this case, we have the following

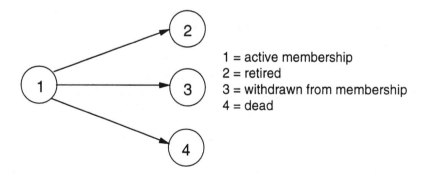

1 = active membership
2 = retired
3 = withdrawn from membership
4 = dead

Fig. 1.6 A four-state model with three causes of 'decrement'.

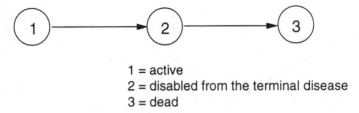

1 = active
2 = disabled from the terminal disease
3 = dead

Fig. 1.7 A further three-state model.

three states: 'active', 'disabled from the terminal disease', 'dead'. If the probability of death among those not suffering from the disease is deemed to be sufficiently small for it to be ignored in the model, we can consider only the transitions depicted in Fig. 1.7.

Example 8

A more complicated structure than Example 5 is needed in order to represent an **annuity benefit in the case of permanent and total disability**. In this case, the death of the disabled insured must be considered, and then transition $2 \to 3$ must be added to the three-state model. The resulting graph is depicted in Fig. 1.8.

Let n denote the policy term; assume that the annuity is payable if the disability inception time belongs to the interval $[0, n)$. Let r denote the stopping time (from policy issue) of the annuity payment, $r \geq n$; for example, if x is the entry age and ξ is the retirement age, then we can assume $r = \xi - x$. (More general and more realistic policy conditions will be considered in Chapter 3.) The annuity benefit is assumed to be paid continuously at a rate b. Premiums are assumed to be paid continuously at a rate p over $[0, n)$ when the contract stays in state 1 (i.e. the insured is

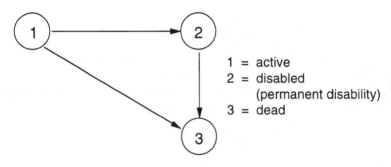

1 = active
2 = disabled
    (permanent disability)
3 = dead

Fig. 1.8 A three-state model with two causes of 'decrement' and with a second-order decrement.

active). Hence, the benefit and premium functions are as follows:

$$b_2(t) = \begin{cases} b & \text{if } 0 < t < r \\ 0 & \text{if } t \geq r \end{cases}$$

$$p_1(t) = \begin{cases} p & \text{if } 0 \leq t < n \\ 0 & \text{if } t \geq n \end{cases}.$$

Remark
Note that transitions $1 \to 2$ and $1 \to 3$, in Example 4, correspond to two 'causes of decrements', according to the traditional actuarial terminology (or 'competing risks', according to the statistical terminology); actually, from a collective point of view, transitions $1 \to 2$ and $1 \to 3$ correspond to decrements in the number of lives belonging to the 'group of alives'. An analogous interpretation holds with reference to Examples 5 and 6. As regards Example 8, transition $2 \to 3$ corresponds to a 'second-order cause of decrement', in the sense that it denotes a decremental factor pertaining to a group (the disabled lives) which, in its turn, has originated by decrement from the 'initial' group (the active lives).

Example 9

Let us generalize Example 8, considering an **annuity benefit in the case of total disability**; thus, the permanent character of the disability is not required. Hence, we have to consider the possibility of recovery, and then the transition $2 \to 1$ must be added to the three-state model depicted in Fig. 1.8. The resulting model is illustrated by Fig. 1.9. If we assume that the policy conditions are as described in Example 6 (as far as policy term, stopping time and premium payment are concerned), then the benefit and premium functions are the same as in Example 6.

Remark
The models presented in Examples 1 to 8 contain only strictly transient states and absorbing states, whilst the model discussed in Example 9 contains (non-strictly)

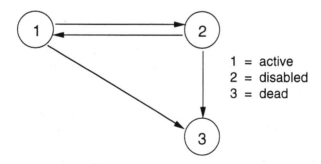

1 = active
2 = disabled
3 = dead

Fig. 1.9 A more general three-state model relating to disability benefits.

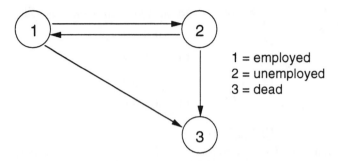

Fig. 1.10 A three-state model relating to an annuity benefit in the case of unemployment.

transient states (states 1 and 2) and an absorbing state (state 3). In section 1.4 we shall consider the difficulties which arise in calculation procedures in the presence of (non-strictly) transient states.

Example 10

Example 9 and the model depicted in Fig. 1.9 can also be adapted to refer to an annuity benefit in the case of unemployment (see Fig. 1.10).

Example 10a

A simplified version of the model depicted in Fig. 1.10 is widely used for practical calculations in respect of annuity benefits paid in the case of unemployment. The model is illustrated in Fig. 1.11, corresponding to Fig. 1.10 but with state 3 omitted completely. This adaptation may be made because the age range covered by such insurance contracts is characterized by low probabilities of death relative to the probabilities of moving from state 1 to state 2 or from state 2 to state 1, or because the financial effects of death may be small relative to that of unemployment.

Example 11

States and transitions can be used as a starting point also to represent insurance contracts with last-survivor benefits. Let us consider a **reversionary**

Fig. 1.11 A simplified two-state model relating to an annuity benefit in the case of unemployment.

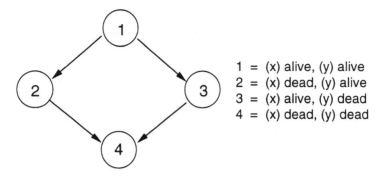

Fig. 1.12 A four-state model.

annuity (i.e. a **widow's pension**). Two lives are involved, say (x) and (y), and then four states must be considered. It is assumed that (x) and (y) cannot die simultaneously. States and transitions are illustrated in Fig. 1.12.

The policy provides a continuous life annuity to (y) at a rate b after the death of (x). Premiums are assumed to be paid continuously at a rate p over $[0, n)$, while both (x) and (y) are alive. Hence, the benefit and premium functions are as follows:

$$b_2(t) = b \quad (t > 0)$$

$$p_1(t) = \begin{cases} p & \text{if } 0 \leq t < n \\ 0 & \text{if } t \geq n \end{cases}.$$

1.4 THE TIME-CONTINUOUS MARKOV MODEL

1.4.1 Some preliminary ideas

It is well known that actuarial evaluations (which are needed in order to calculate single premiums, periodic premiums, mathematical reserves, etc.) include:

1. the calculation of present values;
2. the calculation of expected values.

Remark
More precisely, we will calculate the expected value of the present value of a stream of payments over time. The reader is referred to Section 1.9, where the calculation of expected present values ('actuarial' values) is discussed.

To perform calculation 1, we need a 'financial structure'. Here, the simplest (and very common in practice) structure is assumed, i.e. the compound interest model with a non-stochastic, constant force of interest δ.

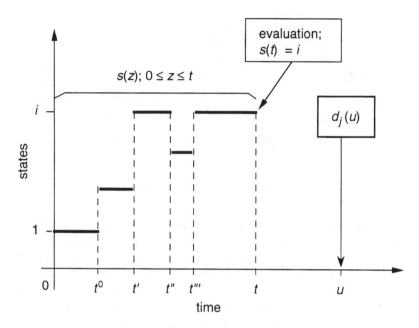

Fig. 1.13 The path of an insured risk.

Hence, the discount function is equal to $e^{-\delta t}$. Let v denote the annual discount factor; thus, $v = e^{-\delta}$.

In order to perform calculation 2, a probabilistic structure must be assumed, pertaining to the events which are relevant to the path of the insured risk and then to the outflow of benefits, and to the inflow of premiums as well. Many possible approaches to stochastic modelling are in principle available; in practice, however, our choice is often subject to constraints imposed by the available data. We shall now begin by discussing a simple example.

Consider Fig. 1.13, in which the path of an insured risk is represented. Suppose that, at time t, with $S(t) = i$, we wish to evaluate the benefit $d_j(u)$ (a lump sum to be paid out if the risk at time u is in state j), i.e. to calculate its expected present value. At time t the story of the insured risk, since the policy issue is known, so that the following conditional expected present value can be calculated:

$$d_j(u)v^{u-t}\Pr\{S(u) = j \mid S(z) = s(z); 0 \leq z \leq t\}, \tag{1.3}$$

where, in particular, $s(t) = i$.

It is unlikely that the probability which appears in equation (1.3) can be evaluated for any conditioning path $s(z)$ (and in actuarial practice this is due to scanty data), so that we have to resort to simpler evaluation schemes. 'Simpler' probabilities can be built up by replacing the conditioning event, and in particular by summarizing the information provided by the path of the insured risk. In other (and more formal) terms, convenient functions of $s(z)$, $0 \leq z \leq t$, shall be defined. Instead of

$$\Pr\{S(u) = j \mid S(z) = s(z); 0 \leq z \leq t\},$$

the following probabilities could be used in calculating the conditional expected present value of the benefit:

$$\Pr\{S(u) = j \mid S(t) = i; \text{ time spent in state } i \text{ since latest transition to } i\} \tag{1.4a}$$

(in the example of Fig. 1.13: time $= t - t'''$);

$$\Pr\{S(u) = j \mid S(t) = i; \text{ time spent in state } i \text{ in } [0, t]\} \tag{1.4b}$$

(in the example of Fig. 1.13: time $= (t'' - t') + (t - t''')$);

$$\Pr\{S(u) = j \mid S(t) = i; \text{ number of stays in state } i \text{ in } [0, t]\} \tag{1.4c}$$

(in the example of Fig. 1.13: number $= 2$);

$$\Pr\{S(u) = j \mid S(t) = i\}. \tag{1.4d}$$

Moreover, information concerning states other than state i could be used in defining the conditioning.

It is self-evident that probability (1.4d) is the 'simplest' one out of the set of conditional probabilities presented above. As we will see below, the use of probability (1.4d) comes from the Markov assumption. Moreover, we will see that probability (1.4a) can be dealt with in the framework of a semi-Markovian model; finally, we will discuss the use of probabilities (1.4a) and (1.4c) within a Markov model, upon redefinition of the state space.

1.4.2 The Markov property

Consider a time-continuous stochastic process $\{S(t); t \geq 0\}$, with a finite (or denumerably infinite) state space $\mathscr{S}$. We say that $\{S(t); t \geq 0\}$ is a **time-continuous Markov chain** if, for any n and each finite set of times $(0 \leq)\ t_0 < \cdots < t_{n-1} < t_n < u$ and corresponding set of states $i_0, \ldots, i_{n-1}, i_n, j$ in $\mathscr{S}$ with

$$\Pr\{S(t_0) = i_0 \wedge \cdots \wedge S(t_{n-1}) = i_{n-1} \wedge S(t_n) = i_n \wedge S(u) = j\} > 0,$$

the following property (the so-called Markov property) is satisfied:

$$\Pr\{S(u) = j \mid S(t_0) = i_0 \wedge \cdots \wedge S(t_{n-1}) = i_{n-1} \wedge S(t_n) = i_n\}$$
$$= \Pr\{S(u) = j \mid S(t_n) = i_n\}. \tag{1.5}$$

Thus, it is assumed that the conditional probability on the left-hand side of probability (1.5) depends only on the 'most recent' information $S(t_n) = i_n$ and is independent of the path before t_n.

Remark 1
From probability (1.5), conditional probabilities of finite-dimensional events can be immediately derived. For instance, we have, for $w > u > t_n$:

$$\Pr\{S(u) = j \wedge S(w) = k \mid S(t_0) = i_0 \wedge \cdots \wedge S(t_{n-1}) = i_{n-1} \wedge S(t_n) = i_n\}$$
$$= \Pr\{S(u) = j \mid S(t_0) = i_0 \wedge \cdots \wedge S(t_{n-1}) = i_{n-1} \wedge S(t_n) = i_n\}$$
$$\times \Pr\{S(w) = k \mid S(t_0) = i_0 \wedge \cdots \wedge S(t_{n-1}) = i_{n-1} \wedge S(t_n) = i_n \wedge S(u) = j\}$$
$$= \Pr\{S(u) = j \mid S(t_n) = i_n\} \Pr\{S(w) = k \mid S(u) = j\}. \tag{1.6}$$

Moreover, according to the axiomatic approach which has been proposed by Kolmogorov in order to handle time-continuous stochastic processes, if all the finite-dimensional conditional probabilities are specified or can be calculated, then the 'complete knowledge' of the process is achieved. Thus, the conditional probabilities of all future paths can be calculated. A very important example of such probabilities is given by:

$$\Pr\{S(z) = h \text{ for all } z \in [u, w] \mid S(t) = i\} \quad (t \leq u < w). \tag{1.7}$$

Remark 2
The Markov property as defined above (see probability (1.5)) has been expressed in terms of finite-dimensional conditioning events. Such an approach can be found in the theory of stochastic processes as well as in the actuarial field (for example, see Wolthuis, 1994). Nevertheless, an alternative approach can be adopted, that is expressing the Markov property in terms of conditioning paths $S(z) = s(z)$, $0 \leq z \leq t$, and this is likely to be very natural when time-continuous stochastic processes are concerned. Following this approach, we say that $\{S(t): t \geq 0\}$ is a time-continuous Markov chain if for all $\tau, u, 0 \leq \tau < u$, and all integers $i, j, s(z)$, $(0 \leq z < \tau)$ in $\mathcal{S}$, such that

$$\Pr\{[S(z) = s(z), 0 \leq z < \tau] \wedge [S(\tau) = i] \wedge [S(u) = j]\} > 0$$

the following property is satisfied:

$$\Pr\{S(u) = j \mid [S(z) = s(z), 0 \leq z < \tau] \wedge [S(\tau) = i]\}$$
$$= \Pr\{S(u) = j \mid S(\tau) = i\}. \tag{1.8}$$

Actually, this definition is sometimes adopted in the mathematics of stochastic processes (for example, see Ross, 1983), and in actuarial applications as well (for instance, see Jones, 1993). Let us briefly comment on the two definitions, from a purely intuitive point of view. In order to compare (1.8) with (1.5), let $\tau = t_n$, $i = i_n$. Equation (1.8) must be satisfied for any path $s(z)$; hence equation (1.5) is satisfied for any finite set of times and related states (each set being determined by some path $s(z)$). Roughly speaking, each conditioning event in (1.5) is 'less informative' than the corresponding conditioning event in equation (1.8). Thus, a stochastic process satisfying property (1.8) also satisfies (1.5). The converse is not self-evident. Intuitively, we can understand that the converse is also true, at least for processes with a finite state space. Actually, in this case any path (of the type specified in (1.8) can be expressed by the 'finite information', given by the set of transition times and the set of states visited by the path.

The conditional probabilities $\Pr\{S(u) = j \mid S(t) = i\}$, for $0 \leq t < u$ and $i, j \in \mathcal{S}$, are called **transition probabilities** and are denoted by:

$$P_{ij}(t, u) = \Pr\{S(u) = j \mid S(t) = i\}. \tag{1.9}$$

For $t \geq 0$ we define:

$$P_{ij}(t, t) = \delta_{ij}, \tag{1.9'}$$

where δ_{ij} denotes the Kronecker delta, which is equal to 0 for $i \neq j$ and equal to 1 for $i = j$. Then, the conditional probabilities $P_{ij}(t, u)$ are defined for $0 \leq t \leq u$.

If, for all t, u such that $0 \leq t \leq u$ and all i, j in $\mathcal{S}$, the transition probability $P_{ij}(t, u)$ depends only on $u - t$ and not on t and u individually, then we say that the process is **time-homogeneous**; in this case the transition probabilities can be simply denoted by $P_{ij}(u - t)$. Otherwise, we say that the process is **time-inhomogeneous**. In actuarial applications, the seniority parameter is usually age related, and the states are often health related or have a demographic meaning, so that it is convenient to assume that the transition probabilities for any fixed period of time $u - t$ vary in time. Hence, in most actuarial applications, time-inhomogeneous processes are concerned.

The transition probabilities satisfy the following properties (i.e. the **probability conditions**):

$$0 \leq P_{ij}(t, u) \leq 1, \quad \text{for all } i, j; \quad 0 \leq t \leq u \tag{1.10a}$$

$$\sum_{j \in \mathcal{S}} P_{ij}(t, u) = 1, \quad \text{for all } i; \quad 0 \leq t \leq u. \tag{1.10b}$$

Moreover, we shall be interested in the following probabilities (often called **occupancy probabilities**):

$$P_{\underline{ii}}(t, u) = \Pr\{S(z) = i \text{ for all } z \in [t, u] \mid S(t) = i\}. \tag{1.11}$$

The transition probabilities satisfy the **Chapman–Kolmogorov equations**

$$P_{ij}(t, u) = \sum_{k \in \mathcal{S}} P_{ik}(t, w) P_{kj}(w, u) \quad (t \leq w \leq u). \tag{1.12}$$

This equation expresses the fact that a path which starts in state i at time t and is in state j at time u visits some state k at an arbitrary intermediate time w. Equation (1.12) can be proved immediately by using the

Markov property (in particular, using definition (1.5)). We have:

$$
\begin{aligned}
P_{ij}(t,u) &= \Pr\{S(u) = j \mid S(t) = i\} \\
&= \sum_{k \in \mathscr{S}} \Pr\{S(u) = j \wedge S(w) = k \mid S(t) = i\} \\
&= \sum_{k \in \mathscr{S}} \Pr\{S(w) = k \mid S(t) = i\} \Pr\{S(u) = j \mid S(t) = i \wedge S(w) = k\} \\
&= \sum_{k \in \mathscr{S}} \Pr\{S(w) = k \mid S(t) = i\} \Pr\{S(u) = j \mid S(w) = k\} \\
&= \sum_{k \in \mathscr{S}} P_{ik}(t,w) P_{kj}(w,u).
\end{aligned}
$$

The occupancy probabilities satisfy the following relation:

$$
P_{\underline{ii}}(t,u) = P_{\underline{ii}}(t,w) P_{\underline{ii}}(w,u) \quad (t \le w \le u). \tag{1.13}
$$

Relation (1.13) can be easily proved by using the Markov property as defined by equation (1.8). We have:

$$
\begin{aligned}
P_{\underline{ii}}(t,u) &= \Pr\{S(z) = i \text{ for all } z \in [t,u] \mid S(t) = i\} \\
&= \Pr\{(S(z) = i \text{ for all } z \in [t,w]) \wedge (S(z) = i \\
&\qquad\qquad\qquad\qquad \text{for all } z \in [w,u]) \mid S(t) = i\} \\
&= \Pr\{(S(z) = i \text{ for all } z \in [t,w]) \mid S(t) = i\} \\
&\qquad \times \Pr\{(S(z) = i \text{ for all } z \in [w,u]) \mid S(z) = i \text{ for all } z \in [t,w]\} \\
&= \Pr\{(S(z) = i \text{ for all } z \in [t,w]) \mid S(t) = i\} \\
&\qquad \times \Pr\{(S(z) = i \text{ for all } z \in [w,u]) \mid S(w) = i\} \\
&= P_{\underline{ii}}(t,w) P_{\underline{ii}}(w,u).
\end{aligned}
$$

Transition probabilities and occupancy probabilities can be used as a tool to formally label the states of a Markov process. In particular (see also section 1.1), we say that

- state i is an **absorbing state** if

$$
P_{\underline{ii}}(t,u) = 1 \quad (0 \le t \le u); \tag{1.14}
$$

- state i is a **transient state** if

$$
P_{ii}(t,+\infty) = 0 \quad (t \ge 0); \tag{1.15}
$$

- state i is a **strictly transient state** if

$$
P_{ii}(t,u) = P_{\underline{ii}}(t,u) < 1 \quad (0 \le t \le u). \tag{1.16}
$$

1.4.3 Transition intensities

The transition intensities are defined as:

$$\mu_{ij}(t) = \lim_{u \to t} \frac{P_{ij}(t, u)}{u - t}. \tag{1.17}$$

These limits are assumed to exist for all relevant t and $i \neq j$, and the intensities are assumed to be integrable on compact intervals. In practice, we are usually interested in continuous (and bounded) intensities and in piecewise constant intensities (with a finite number of jumps). If the Markov process is assumed to be time-homogeneous, then we have:

$$\mu_{ij}(t) = \lim_{u \to t} \frac{P_{ij}(u - t)}{u - t} = \mu_{ij}, \tag{1.17'}$$

thus the transition intensities are constant functions.

Note that the expression $\mu_{ij}(t)\, dt$ can be interpreted as the conditional probability that a transition from state i into state j occurs over the infinitesimal interval $[t, t + dt)$ given that the risk is in state i at time t.

Moreover, let us define the (total) **intensity of decrement** from state i:

$$\mu_i(t) = \sum_{j: j \neq i} \mu_{ij}(t). \tag{1.18}$$

From equation (1.17) we have:

$$\mu_i(t) = \sum_{j: j \neq i} \lim_{u \to t} \frac{P_{ij}(t, u)}{u - t}$$

$$= \lim_{u \to t} \frac{\sum_{j: j \neq i} P_{ij}(t, u)}{u - t}$$

$$= \lim_{u \to t} \frac{1 - P_{ii}(t, u)}{u - t}. \tag{1.18}$$

Note that the expression $\mu_i(t)\, dt$ can be interpreted as the conditional probability of leaving state i over the infinitesimal interval $[t, t + dt)$ given that the risk is in state i at time t.

Two sets of differential equations for the transition probabilities can be derived. First, consider the **Kolmogorov forward differential equations**:

$$\frac{d}{dt} P_{ij}(z, t) = \sum_{k: k \neq j} P_{ik}(z, t)\mu_{kj}(t) - P_{ij}(z, t)\mu_j(t) \tag{1.19}$$

for all states i, j and times t, $0 \leq z \leq t$. The boundary conditions are given by $P_{ij}(z, z) = \delta_{ij}$ (where δ_{ij} is the Kronecker delta). Equation (1.19) can be easily derived, starting from the Chapman–Kolmogorov equations (see equation (1.12)):

$$P_{ij}(z, t + \Delta t) = \sum_{k: k \neq j} P_{ik}(z, t) P_{kj}(t, t + \Delta t) + P_{ij}(z, t) P_{jj}(t, t + \Delta t);$$

then:

$$\frac{P_{ij}(z, t + \Delta t) - P_{ij}(z, t)}{\Delta t} = \sum_{k: k \neq j} P_{ik}(z, t) \frac{P_{kj}(t, t + \Delta t)}{\Delta t}$$

$$+ P_{ij}(z, t) \frac{P_{jj}(t, t + \Delta t) - 1}{\Delta t};$$

since

$$1 - P_{jj}(t, t + \Delta t) = \sum_{k: k \neq j} P_{jk}(t, t + \Delta t)$$

we have:

$$\frac{P_{ij}(z, t + \Delta t) - P_{ij}(z, t)}{\Delta t} = \sum_{k: k \neq j} P_{ik}(z, t) \frac{P_{kj}(t, t + \Delta t)}{\Delta t}$$

$$- P_{ij}(z, t) \sum_{k: k \neq j} \frac{P_{jk}(t, t + \Delta t)}{\Delta t};$$

finally, as $\Delta t \to 0^+$, we get equation (1.19).

Equation (1.19) can be written in differential form, in which it is easier to interpret:

$$dP_{ij}(z, t) = \sum_{k: k \neq j} P_{ik}(z, t) \mu_{kj}(t)\, dt - P_{ij}(z, t) \mu_j(t)\, dt. \tag{1.19'}$$

We can read equation (1.19') as follows. In all transition probabilities, the risk is in state i at time z; the left-hand side represents the change in the probability of entering state j over $[t, t + dt)$; the right-hand side represents the probability of entering state j starting from any state k, $k \neq j$, less the probability of leaving state j, both the probabilities referring to the interval $[t, t + dt)$.

The **Kolmogorov backward differential equations** are as follows:

$$\frac{d}{dz} P_{ij}(z, t) = P_{ij}(z, t) \mu_i(z) - \sum_{k: k \neq i} P_{kj}(z, t) \mu_{ik}(z) \tag{1.20}$$

with boundary conditions $P_{ij}(t, t) = \delta_{ij}$ (where δ_{ij} is the Kronecker delta). For brevity, we omit the construction and interpretation of equation (1.20).

For the occupancy probabilities, the following equation can be derived:

$$\frac{d}{dt} P_{\underline{ii}}(z, t) = -P_{\underline{ii}}(z, t) \mu_i(t); \tag{1.21}$$

the boundary condition is $P_{\underline{ii}}(z,z) = 1$. The derivation of equation (1.21) is immediate. From relation (1.13) we have:

$$P_{\underline{ii}}(z, t + \Delta t) = P_{\underline{ii}}(z, t)P_{\underline{ii}}(t, t + \Delta t);$$

then:

$$\frac{P_{\underline{ii}}(z, t + \Delta t) - P_{\underline{ii}}(z, t)}{\Delta t} = \frac{P_{\underline{ii}}(z, t)[P_{\underline{ii}}(t, t + \Delta t) - 1]}{\Delta t};$$

the probability of two or more transitions in the interval $[t, t + \Delta t)$ is $o(\Delta t)$, a quantity such that

$$\lim_{\Delta t \to 0^+} \frac{o(\Delta t)}{\Delta t} = 0;$$

hence:

$$P_{\underline{ii}}(t, t + \Delta t) + o(\Delta t) = P_{ii}(t, t + \Delta t);$$

then we can write:

$$\frac{P_{\underline{ii}}(z, t + \Delta t) - P_{\underline{ii}}(z, t)}{\Delta t} = -P_{\underline{ii}}(z, t)\frac{1 - P_{ii}(t, t + \Delta t) + o(\Delta t)}{\Delta t}.$$

Letting $\Delta t \to 0^+$ we finally obtain the differential equation (1.21).

In the so-called **transition intensity approach** (TIA), it is assumed that the transition intensities are assigned. From the intensities, via differential equations, the transition probabilities can (at least) in principle be derived. In the actuarial practice of insurances of the person, the intensities should be estimated from statistical data concerning mortality, disability, recovery, etc. As far as disability benefits are concerned, the reader can refer to Chapter 4, where the estimation of the relevant intensities from observed statistical data is discussed.

It is beyond the scope of this book to discuss the existence of solutions for the transition probabilities. We simply assert that under general conditions for the transition intensities (and since we are working with a finite state space), each of the sets of simultaneous differential equations (1.19) and (1.20) uniquely determine the transition probabilities $P_{ij}(z, t)$, and that these transition probabilities satisfy the probability conditions (1.10) and the Chapman–Kolmogorov equations (1.12).

As far as the occupancy probabilities are concerned, the solution is trivial. From equation (1.21) we get:

$$\frac{\mathrm{d}}{\mathrm{d}t} \ln P_{\underline{ii}}(z, t) = -\mu_i(t)$$

whence, using the boundary condition $P_{\underline{ii}}(z, z) = 1$, we obtain:

$$P_{\underline{ii}}(z, t) = \exp\left[-\int_z^t \mu_i(u)\,\mathrm{d}u\right].$$

On the contrary, the calculation of the transition probabilities may in general be a difficult integration problem. In the next section we will examine several particular problems, in some of which the solution is easily achieved.

Sometimes, the set of transition intensities is denoted, in matrix form, by:

$$\mathbf{M}(t) = ||\mu_{ij}(t)||;$$

by definition, we assume $\mu_{ii}(t) = -\mu_i(t)$ for all i in $\mathscr{S}$; of course, $\mu_{ij}(t)$ is identically equal to 0 if (i,j) is not in $\mathscr{T}$. Note that, according to the TIA, the probabilistic structure is completely defined by $\mathbf{M}(t)$. Then, we can say that, in the Markov context, the multiple state model $(\mathscr{S}, \mathscr{T})$ is valued by $\mathbf{M}(t)$.

1.5 EXAMPLES

The examples that we discuss in this section are as presented in section 1.3. Here we will group the examples according to the structure of the graph which describes the states and transitions. In particular, we will verify that the degree of complexity in the solution of the set of differential equations is strictly related to the structure of the graph itself.

1.5.1 Examples 1, 2, 3

In these examples the randomness only comes from the lifetime of the insured; causes of death are not separately considered. State 1 is a strictly transient state and state 2 is an absorbing one (see Fig. 1.3). According to the TIA, the transition intensity $\mu_{12}(t)$, i.e. the **intensity of mortality**, must be assigned. Of course $\mu_{21}(t)$ is identically equal to zero. Since state 2 is an absorbing state, we have:

$$P_{\underline{11}}(z,t) = P_{11}(z,t); \quad 0 \le z \le t. \tag{1.22}$$

The set of (forward) differential equations (see equation (1.19)) is as follows:

$$\frac{d}{dt} P_{11}(z,t) = - P_{11}(z,t)\mu_{12}(t) \tag{1.23a}$$

$$\frac{d}{dt} P_{12}(z,t) = P_{11}(z,t)\mu_{12}(t). \tag{1.23b}$$

From equation (1.23a), using the boundary condition $P_{11}(z,z) = 1$, we get:

$$P_{11}(z,t) = \exp\left[-\int_z^t \mu_{12}(u)\, du\right]. \tag{1.24a}$$

The transition probability $P_{12}(z,t)$ can be found immediately using the probability condition:

$$P_{12}(z,t) = 1 - P_{11}(z,t). \qquad (1.24b)$$

It is interesting to outline the connections between the notation used to describe the Markov process and the traditional actuarial notation. The traditional notation for the intensity of mortality is μ_{x+t}, where $x+t$ denotes the attained age and x the age at policy issue, i.e. at time 0. In the Markov process notation age x does not appear, and this is an advantage when more than one insured is concerned (see for instance Example 11 in section 1.3 and in this section). Nevertheless, age has an obvious relevance in determining the mortality of an individual. Therefore, the notation $\mu_{12}(t)$ must be interpreted in one of the following ways.

1. $\mu_{12}(t) = \mu_{x+t}$. In this case the transition intensity $\mu_{12}(t)$ is simply a 'part' (a 'restriction', in mathematical terms) of the intensity of mortality μ_{x+t} (which can be given by one of the classical models, such as the models of Gompertz, Makeham, Thiele, etc.). The entry age x is in this case an 'offset' parameter. Thus the values of $\mu_{12}(t)$ are the values of μ_{x+t} for $t \geq 0$. Since μ_{x+t} is a function of the attained age only, the model is called **aggregate**.
2. $\mu_{12}(t) = \mu_{[x]+t}$, where $\mu_{[x]+t}$, according to the actuarial notation, denotes a function of the two variables x and t separately. Hence we have a different function for each entry age x, and this can reflect the lower mortality during the first policy years for standard risks who have passed a medical ascertainment at policy issue, i.e. a 'selection'. For this reason, the model is usually called **select**, and more precisely **issue-select** (a different type of selection shall be considered later).

According to the traditional notation and considering for simplicity an aggregate model, we have:

$$P_{11}(z,t) = {}_{t-z}p_{x+z}; \qquad (1.25)$$

$$P_{12}(z,t) = {}_{t-z}q_{x+z}; \qquad (1.26)$$

$$P_{11}(z,t)\mu_1(t)\,dt = {}_{t-z}p_{x+z}\mu_{x+t}\,dt = {}_{t-z|dt}q_{x+z}. \qquad (1.27)$$

Moreover, note that the expected residual lifetime at age $x+z$, denoted by $\bar{e}_{x+z}$, can be expressed as follows:

$$\bar{e}_{x+z} = \int_z^{+\infty} {}_{t-z}p_{x+z}\,dt = \int_z^{+\infty} P_{11}(z,t)\,dt. \qquad (1.28)$$

1.5.2 Examples 4, 5

In these examples a three-state model is concerned, with a strictly transient state and two absorbing states. In both cases, two transition intensities are involved, i.e. $\mu_{12}(t)$ and $\mu_{13}(t)$. In Example 4 their meaning is:

$$\mu_{12}(t) = \text{intensity of non-accidental mortality}$$
$$\mu_{13}(t) = \text{intensity of accidental mortality};$$

while in Example 5 the meaning is as follows:

$$\mu_{12}(t) = \text{intensity of (total and permanent) disability}$$
$$\mu_{13}(t) = \text{intensity of mortality for active lives}.$$

Since states 2 and 3 are absorbing states, we have

$$P_{\underline{11}}(z,t) = P_{11}(z,t); \quad 0 \le z \le t.$$

In both cases, the set of simultaneous differential equations is as follows:

$$\frac{d}{dt}P_{11}(z,t) = -P_{11}(z,t)(\mu_{12}(t) + \mu_{13}(t)) \tag{1.29a}$$

$$\frac{d}{dt}P_{12}(z,t) = P_{11}(z,t)\mu_{12}(t) \tag{1.29b}$$

$$\frac{d}{dt}P_{13}(z,t) = P_{11}(z,t)\mu_{13}(t). \tag{1.29c}$$

From equation (1.29a), using the initial condition $P_{11}(z,z) = 1$, we obtain:

$$P_{11}(z,t) = \exp\left[-\int_z^t (\mu_{12}(u) + \mu_{13}(u))\,du\right]. \tag{1.30a}$$

Then, using the initial condition $P_{12}(z,z) = 0$, we have:

$$P_{12}(z,t) = \int_z^t P_{11}(z,u)\mu_{12}(u)\,du \tag{1.30b}$$

and finally:

$$P_{13}(z,t) = 1 - P_{11}(z,t) - P_{12}(z,t). \tag{1.30c}$$

1.5.3 Example 6

Here, a four-state model is under consideration, with a strictly transient state and three absorbing states. Three transition intensities are involved:

$$\mu_{12}(t) = \text{intensity of retirement};$$
$$\mu_{13}(t) = \text{intensity of withdrawal};$$
$$\mu_{14}(t) = \text{intensity of mortality for active lives}.$$

The set of simultaneous differential equations is as follows:

$$\frac{d}{dt}P_{11}(z,t) = -P_{11}(z,t)(\mu_{12}(t) + \mu_{13}(t) + \mu_{14}(t)) \qquad (1.31a)$$

$$\frac{d}{dt}P_{12}(z,t) = P_{11}(z,t)\mu_{12}(t) \qquad (1.31b)$$

$$\frac{d}{dt}P_{13}(z,t) = P_{11}(z,t)\mu_{13}(t) \qquad (1.31c)$$

$$\frac{d}{dt}P_{14}(z,t) = P_{11}(z,t)\mu_{14}(t) \qquad (1.31d)$$

which can be integrated in a similar manner to Examples 4 and 5 in section 1.5.2 above.

1.5.4 Example 7

Here, a three-state model is under consideration, with two strictly transient states and one absorbing state. The transition intensities involved are:

$\mu_{12}(t)$ = intensity of incidence of terminal disease

$\mu_{23}(t)$ = intensity of mortality of terminal disease.

The set of simultaneous differential equations is as follows:

$$\frac{d}{dt}P_{11}(z,t) = -P_{11}(z,t)\mu_{12}(t) \qquad (1.32a)$$

$$\frac{d}{dt}P_{12}(z,t) = P_{11}(z,t)\mu_{12}(t) - P_{12}(z,t)\mu_{23}(t) \qquad (1.32b)$$

$$\frac{d}{dt}P_{13}(z,t) = P_{12}(z,t)\mu_{23}(t) \qquad (1.32c)$$

$$\frac{d}{dt}P_{22}(z,t) = -P_{22}(z,t)\mu_{23}(t) \qquad (1.32d)$$

$$\frac{d}{dt}P_{23}(z,t) = P_{22}(z,t)\mu_{23}(t). \qquad (1.32e)$$

1.5.5 Example 8

The situation seems to be similar to that considered in Examples 4 and 5, but the transition $2 \rightarrow 3$ must be added to the three-state model in order to express the mortality of disabled lives. Then, three transition intensities are involved:

$\mu_{12}(t)$ = intensity of (total and permanent) disability

$\mu_{13}(t)$ = intensity of mortality for active lives

$\mu_{23}(t)$ = intensity of mortality for disabled lives.

Note that state 2 is now a strictly transient state. The set of simultaneous differential equations is as follows:

$$\frac{d}{dt}P_{11}(z,t) = -P_{11}(z,t)(\mu_{12}(t) + \mu_{13}(t)) \tag{1.33a}$$

$$\frac{d}{dt}P_{12}(z,t) = P_{11}(z,t)\mu_{12}(t) - P_{12}(z,t)\mu_{23}(t) \tag{1.33b}$$

$$\frac{d}{dt}P_{13}(z,t) = P_{11}(z,t)\mu_{13}(t) + P_{12}(z,t)\mu_{23}(t) \tag{1.33c}$$

$$\frac{d}{dt}P_{23}(z,t) = P_{22}(z,t)\mu_{23}(t) \tag{1.33d}$$

$$\frac{d}{dt}P_{22}(z,t) = -P_{22}(z,t)\mu_{23}(t). \tag{1.33e}$$

From equations (1.33a) and (1.33e), using the boundary conditions

$$P_{11}(z,z) = P_{22}(z,z) = 1,$$

we respectively obtain:

$$P_{11}(z,t) = \exp\left[-\int_z^t (\mu_{12}(u) + \mu_{13}(u))\,du\right] \tag{1.34a}$$

$$P_{22}(z,t) = \exp\left[-\int_z^t \mu_{23}(u)\,du\right] \tag{1.34e}$$

whence

$$P_{23}(z,t) = 1 - P_{22}(z,t). \tag{1.34d}$$

Then, using the relevant boundary conditions and after some manipulations, we obtain:

$$P_{12}(z,t) = \int_z^t P_{11}(z,u)\mu_{12}(u)P_{22}(u,t)\,du. \tag{1.34b}$$

Finally

$$P_{13}(z,t) = 1 - P_{11}(z,u) - P_{12}(z,u). \tag{1.34c}$$

As states 1 and 2 are strictly transient states, we have:

$$P_{\underline{1}1}(z,t) = P_{11}(z,t) \tag{1.35a}$$

$$P_{\underline{2}2}(z,t) = P_{22}(z,t). \tag{1.35b}$$

Remark 1

Example 8 discussed above constitutes a first simple but realistic model for disability annuities. The model is used for instance in Ramlau-Hansen (1991), with Makeham-like transition intensities. The intensities are chosen as follows:

$$\mu_{12}(t) = 0.0004 + 10^{0.06(x+t)-5.46}$$

$$\mu_{13}(t) = \mu_{23}(t) = 0.0005 + 10^{0.038(x+t)-4.12}.$$

We can in particular note that:

- mortality for active lives and for disabled lives are not discriminated;
- the model is of the aggregate type, since all the transition intensities depend on attained age $x + t$ only.

Remark 2

It is very important to remark that in Examples 1 to 8 a 'sequential' procedure can be adopted in solving the set of simultaneous equations, that is in finding the transition probabilities. It is rather easy to realize that such an opportunity is allowed by the 'hierarchical' structure of the graph which describes the model (see, for instance, Fig. 1.6), i.e. by the fact that the set of nodes is an ordered set. In Example 9 we will deal with a model based on a non-ordered set of nodes; as a consequence, we will see that a sequential procedure cannot be followed for finding the transition probabilities.

1.5.6 Example 9

Now the transition $2 \to 1$ must be added to the model discussed in Example 8, in order to express the possibility of recovery. We will see that in this case it is much more difficult to express the transition probabilities in terms of the transition intensities. Four transition intensities are now involved:

$$\mu_{12}(t) = \text{intensity of (total) disability}$$

$$\mu_{21}(t) = \text{intensity of recovery}$$

$$\mu_{13}(t) = \text{intensity of mortality for active lives}$$

$$\mu_{23}(t) = \text{intensity of mortality for disabled lives.}$$

The set of simultaneous forward differential equations is as follows:

$$\frac{d}{dt}P_{11}(z,t) = P_{12}(z,t)\mu_{21}(t) - P_{11}(z,t)[\mu_{12}(t) + \mu_{13}(t)] \quad (1.36a)$$

$$\frac{d}{dt}P_{12}(z,t) = P_{11}(z,t)\mu_{12}(t) - P_{12}(z,t)[\mu_{21}(t) + \mu_{23}(t)] \quad (1.36b)$$

$$\frac{d}{dt}P_{13}(z,t) = P_{11}(z,t)\mu_{13}(t) + P_{12}(z,t)\mu_{23}(t) \quad (1.36c)$$

$$\frac{d}{dt}P_{21}(z,t) = P_{22}(z,t)\mu_{21}(t) - P_{21}(z,t)[\mu_{12}(t) + \mu_{13}(t)] \quad (1.36d)$$

$$\frac{d}{dt}P_{22}(z,t) = P_{21}(z,t)\mu_{12}(t) - P_{22}(z,t)[\mu_{21}(t) + \mu_{23}(t)] \qquad (1.36e)$$

$$\frac{d}{dt}P_{23}(z,t) = P_{22}(z,t)\mu_{23}(t) + P_{21}(z,t)\mu_{13}(t). \qquad (1.36f)$$

Note that, unlike Example 8, nodes 1 and 2 are non-strictly transient nodes; hence identities (1.35a) and (1.35b) do not hold. The following differential equations pertain to the occupancy probabilities:

$$\frac{d}{dt}P_{\underline{11}}(z,t) = -P_{\underline{11}}(z,t)[\mu_{12}(t) + \mu_{13}(t)] \qquad (1.37a)$$

$$\frac{d}{dt}P_{\underline{22}}(z,t) = -P_{\underline{22}}(z,t)[\mu_{21}(t) + \mu_{23}(t)]. \qquad (1.37b)$$

The solution of equations (1.37a) and (1.37b) is trivial; we get:

$$P_{\underline{11}}(z,t) = \exp\left[-\int_z^t [\mu_{12}(u) + \mu_{13}(u)]\,du\right] \qquad (1.38a)$$

$$P_{\underline{22}}(z,t) = \exp\left[-\int_z^t [\mu_{21}(u) + \mu_{23}(u)]\,du\right]. \qquad (1.38a)$$

The set of differential equations (1.36a) to (1.36f) can be solved as follows. Equations (1.36a) and (1.36b) constitute a pair of simultaneous differential equations with two unknown functions $P_{11}(z,t)$ and $P_{12}(z,t)$. By differentiation and substitution a second-order differential equation can be obtained for $P_{11}(z,t)$ (or for $P_{12}(z,t)$); this equation can be solved by numerical methods, allowing for the relevant boundary conditions. From the solution for $P_{11}(z,t)$, we can by substitution obtain the solution for $P_{12}(z,t)$. With these two solutions, we can solve the differential equation for $P_{13}(z,t)$, i.e. equation (1.36c). The same procedure can be adopted for the subset of equations (1.36d), (1.36e) and (1.36f).

In order to illustrate the procedure, we shall take equations (1.36a) and (1.36b). From (1.36a) we have:

$$P_{12}(z,t) = \frac{1}{\mu_{21}(t)}\left[\frac{d}{dt}P_{11}(z,t) + P_{11}(z,t)[\mu_{12}(t) + \mu_{13}(t)]\right]; \qquad (1.36a')$$

substituting into equation (1.36b), after some manipulations we obtain:

$$\frac{d^2}{dt^2}P_{11}(z,t) + \left[\frac{d}{dt}\ln\mu_{21}(t) + \mu_{12}(t) + \mu_{13}(t) + \mu_{21}(t) + \mu_{23}(t)\right]\frac{d}{dt}P_{11}(z,t)$$

$$+ \left[\frac{d}{dt}\ln\mu_{21}(t)[\mu_{12}(t) + \mu_{13}(t)] + \frac{d}{dt}\mu_{12}(t) + \frac{d}{dt}\mu_{13}(t) - \mu_{12}(t)\mu_{21}(t)\right.$$

$$\left. + [\mu_{12}(t) + \mu_{13}(t)][\mu_{21}(t) + \mu_{23}(t)]\right]P_{11}(z,t) = 0. \qquad (1.39)$$

The boundary conditions are:

$$P_{11}(z,z) = 1$$

$$\frac{d}{dt}P_{11}(z,t)|_{t=z} = -\mu_{12}(z) - \mu_{13}(z).$$

The second condition is obtained from equation (1.36a) with $P_{12}(z,t) = 0$ when $t = z$.

An interesting alternative expression for probability $P_{12}(z,t)$ is as follows:

$$P_{12}(z,t) = \int_z^t P_{11}(z,u)\mu_{12}(u)P_{\underline{22}}(u,t)\,du. \tag{1.40}$$

It is possible to prove that the right-hand side member of equation (1.40) satisfies the differential equation (1.36b). Equation (1.40) can be easily interpreted by direct reasoning. Moreover, it is interesting to compare it with the analogous equation (1.34b) pertaining to the permanent disability annuity; note that when only permanent disability is considered $P_{\underline{22}} = P_{22}$. Equation (1.40) represents a useful tool in deriving formulae for premium calculations, as we will see in section 1.10.

1.5.7 Example 9 (continued)

Particular assumptions about the transition intensities can lead to substantial simplifications in finding transition probabilities. We now consider some particular cases.

1. Consider the assumption $\mu_{13}(t) = \mu_{23}(t) = \mu(t)$, say, which is often referred to as 'no differential mortality'. In this case we obtain:

$$P_{13}(z,t) = 1 - \exp\left[-\int_z^t \mu(u)\,du\right]. \tag{1.41}$$

This result can be proved as follows. Equation (1.36c) now becomes:

$$\frac{d}{dt}P_{13}(z,t) = \mu(t)[P_{11}(z,t) + P_{12}(z,t)]$$

and then

$$\frac{d}{dt}[1 - P_{13}(z,t)] = -\mu(t)[1 - P_{13}(z,t)]. \tag{1.42}$$

Equation (1.42) can be solved with the boundary condition $P_{13}(z,z) = 0$; so expression (1.41) is obtained.

2. Let us assume that the transition intensities are all constants, independent of age, i.e.

$$\mu_{12}(t) = \mu_{12}, \quad \mu_{13}(t) = \mu_{13}, \quad \mu_{21}(t) = \mu_{21}, \quad \mu_{23}(t) = \mu_{23}.$$

The second-order differential equation for $P_{11}(z,t)$, equation (1.39), then becomes:

$$\frac{d^2}{dt^2}P_{11}(z,t) + (\mu_{12} + \mu_{13} + \mu_{21} + \mu_{23})\frac{d}{dt}P_{11}(z,t)$$

$$+ [(\mu_{12} + \mu_{13})(\mu_{21} + \mu_{23}) - \mu_{12}\mu_{21}]P_{11}(z,t) = 0; \qquad (1.43)$$

with boundary conditions:

$$P_{11}(z,z) = 1 \qquad (1.44)$$

$$\frac{d}{dt}P_{11}(z,t)|_{t=z} = -\mu_{12} - \mu_{13}. \qquad (1.45)$$

A trial solution of the form $P_{11}(z,t) = e^{r(t-z)}$ yields:

$$r^2 + (\mu_{12} + \mu_{13} + \mu_{21} + \mu_{23})r + [(\mu_{12} + \mu_{13})(\mu_{12} + \mu_{23}) - \mu_{12}\mu_{21}] = 0$$

which has solutions r_1 and r_2 where

$$r_1 = \tfrac{1}{2}[-(\mu_{12} + \mu_{13} + \mu_{21} + \mu_{23}) + [(\mu_{12} + \mu_{13} - \mu_{21} - \mu_{23})^2 + 4\mu_{12}\mu_{21}]^{1/2}]$$

$$r_2 = \tfrac{1}{2}[-(\mu_{12} + \mu_{13} + \mu_{21} + \mu_{23}) - [(\mu_{12} + \mu_{13} - \mu_{21} - \mu_{23})^2 + 4\mu_{12}\mu_{21}]^{1/2}].$$

So we have

$$P_{11}(z,t) = A\,e^{r_1(t-z)} + B\,e^{r_2(t-z)} \qquad (1.46)$$

and we can determine A and B from the boundary conditions. From equations (1.44) and (1.45) we respectively have:

$$1 = A + B$$

$$-\mu_{12} - \mu_{13} = Ar_1 + Br_2.$$

Solving this pair of simultaneous equations leads to

$$A = \frac{r_2 + \mu_{12} + \mu_{13}}{r_2 - r_1} \qquad (1.47a)$$

$$B = \frac{r_1 + \mu_{12} + \mu_{13}}{r_1 - r_2}. \qquad (1.47b)$$

Finally, we have:

$$P_{11}(z,t) = \frac{(r_2 + \mu_{12} + \mu_{13})\,e^{r_1(t-z)} - (r_1 + \mu_{12} + \mu_{13})\,e^{r_2(t-z)}}{r_2 - r_1}. \qquad (1.48)$$

Substituting equation (1.48) into (1.36a'), after some manipulations we obtain:

$$P_{12}(z,t) = \frac{\mu_{12}(e^{r_1(t-z)} - e^{r_2(t-z)})}{r_1 - r_2} \qquad (1.49)$$

where r_1 and r_2 are given by the earlier expressions.

3. We now add the assumption of $\mu_{13} = \mu_{23} = \mu$ (that is 'no differential mortality') to the model discussed in case 2. The quadratic equation for r now becomes:

$$r^2 + (\mu_{12} + \mu_{21} + 2\mu)r + \mu^2 + \mu(\mu_{12} + \mu_{21}) = 0,$$

which factorizes as

$$(r + \mu)(r + \mu + \mu_{12} + \mu_{21}) = 0,$$

whence

$$r_1 = -\mu, \quad r_2 = -(\mu + \mu_{12} + \mu_{21}).$$

From equation (1.46) we obtain

$$P_{11}(z, t) = A\,e^{-\mu(t-z)} + B\,e^{-(\mu+\mu_{12}+\mu_{21})(t-z)}. \tag{1.50}$$

From equations (1.47a) and (1.47b) we now have:

$$A = \frac{\mu_{21}}{\mu_{12} + \mu_{21}} \tag{1.51a}$$

$$B = \frac{\mu_{12}}{\mu_{12} + \mu_{21}} \tag{1.51b}$$

so

$$P_{11}(z, t) = \frac{1}{\mu_{12} + \mu_{21}} \left[\mu_{21}\,e^{-\mu(t-z)} + \mu_{12}\,e^{-(\mu+\mu_{12}+\mu_{21})(t-z)}\right]. \tag{1.52}$$

Similarly we obtain

$$P_{12}(z, t) = \frac{\mu_{12}}{\mu_{12} + \mu_{21}} \left[e^{-\mu(t-z)} - e^{-(\mu+\mu_{12}+\mu_{21})(t-z)}\right]. \tag{1.53}$$

By subtraction

$$P_{13}(z, t) = 1 - P_{11}(z, t) - P_{12}(z, t)$$

and we have:

$$P_{13}(z, t) = 1 - e^{-\mu(t-z)}. \tag{1.54}$$

1.5.8 Example 10a

The set of simultaneous differential equations is just a simplified version of equations (1.36) in Example 9, noting the absence of state 3:

$$\frac{\mathrm{d}}{\mathrm{d}t} P_{11}(z, t) = P_{12}(z, t)\mu_{21}(t) - P_{11}(z, t)\mu_{12}(t) \tag{1.55a}$$

$$\frac{\mathrm{d}}{\mathrm{d}t} P_{12}(z, t) = P_{11}(z, t)\mu_{12}(t) - P_{12}(z, t)\mu_{21}(t) \tag{1.55b}$$

$$\frac{d}{dt}P_{21}(z, t) = P_{22}(z, t)\mu_{21}(t) - P_{21}(z, t)\mu_{12}(t) \tag{1.55c}$$

$$\frac{d}{dt}P_{22}(z, t) = P_{21}(z, t)\mu_{12}(t) - P_{22}(z, t)\mu_{21}(t). \tag{1.55d}$$

The discussion of Example 9 is also relevant here.

1.5.9 Example 11

In this four-state model four transition intensities must be considered. All the transition intensities are in this case intensities of mortality. Assume that (x) denotes the husband and (y) denotes the wife.

$\mu_{12}(t)$ = intensity of mortality of (x) (married)

$\mu_{13}(t)$ = intensity of mortality of (y) (married)

$\mu_{24}(t)$ = intensity of mortality of (y) (widow)

$\mu_{34}(t)$ = intensity of mortality of (x) (widower).

It is important to stress that, using the Markov modelling approach, it is quite natural to discriminate between the mortality of a married person and that of a widow (or widower), if the remaining lifetimes seem to be correlated. Hence, we can assume:

$$\mu_{12}(t) \neq \mu_{34}(t); \quad \mu_{13}(t) \neq \mu_{24}(t);$$

in particular, if a positive correlation between remaining lifetimes seems to be reasonable, we should assume:

$$\mu_{12}(t) < \mu_{34}(t); \quad \mu_{13}(t) < \mu_{24}(t).$$

Note that the traditional actuarial approach does not allow for correlation; actually, the mortality functions are simply taken from a male ($\mu_{12}(t) = \mu_{34}(t)$) and a female ($\mu_{13}(t) = \mu_{24}(t)$) population life table.

The set of forward differential equations is as follows:

$$\frac{d}{dt}P_{11}(z, t) = -P_{11}(z, t)[\mu_{12}(t) + \mu_{13}(t)] \tag{1.56a}$$

$$\frac{d}{dt}P_{12}(z, t) = P_{11}(z, t)\mu_{12}(t) - P_{12}(z, t)\mu_{24}(t) \tag{1.56b}$$

$$\frac{d}{dt}P_{13}(z, t) = P_{11}(z, t)\mu_{13}(t) - P_{13}(z, t)\mu_{34}(t) \tag{1.56c}$$

$$\frac{d}{dt}P_{22}(z, t) = -P_{22}(z, t)\mu_{24}(t) \tag{1.56d}$$

$$\frac{d}{dt}P_{33}(z, t) = -P_{33}(z, t)\mu_{34}(t) \tag{1.56e}$$

$$\frac{d}{dt}P_{24}(z,t) = P_{22}(z,t)\mu_{24}(t) \tag{1.56f}$$

$$\frac{d}{dt}P_{34}(z,t) = P_{33}(z,t)\mu_{34}(t). \tag{1.56g}$$

As states 1, 2 and 3 are strictly transient states, we have:

$$P_{\underline{11}}(z,t) = P_{11}(z,t) \tag{1.57a}$$

$$P_{\underline{22}}(z,t) = P_{22}(z,t) \tag{1.57b}$$

$$P_{\underline{33}}(z,t) = P_{33}(z,t). \tag{1.57c}$$

The solution of (1.56a), (1.56d), (1.56e) is trivial. We leave as an exercise the problem of solving the remaining equations; note that the solving procedure can follow a 'sequential' path, thanks to the order in the set of nodes.

Remark
Example 11 above has been discussed numerically by Wolthuis (1994), for the case where

$$\mu_{12}(t) = (1-\alpha)\mu_{x+t}$$

$$\mu_{13}(t) = (1-\alpha)\mu_{y+t}$$

$$\mu_{24}(t) = (1+\alpha)\mu_{y+t}$$

$$\mu_{34}(t) = (1+\alpha)\mu_{x+t}$$

with μ_{x+t} and μ_{y+t} based respectively on male and female population life tables and α a parameter representing the dependence of mortality intensities on current marital status.

1.6 THE SEMI-MARKOV MODEL

1.6.1 Some preliminary ideas

The preceding Markov model assumes that transition intensities (and probabilities) at time t depend (at least explicitly) on the current state at that time only. More realistic (and possibly more complicated) models can be built considering, for instance:

1. the dependence of some intensities (and probabilities) on the age x at policy issue;
2. the dependence of some intensities (and probabilities) on the time spent in the current state since the latest transition to that state (see equation (1.4a) in section 1.4);

3. the dependence of some intensities (and probabilities) on the total time spent in some states since policy issue (a particular case is given by equation (1.4b) in section 1.4).

The consideration of model 1, **duration-since-initiation dependence**, implies the use of **issue-select** transition intensities. As far as the intensity of mortality is concerned, we noted in section 1.5 (Examples 1, 2, 3) that the function $\mu_{12}(t)$ can be interpreted as $\mu_{[x]+t}$, i.e. as a specific function for each entry age x, and this allows for issue-selection in mortality. An analogous interpretation can hold for other transition intensities. For instance, issue-selection in the intensity of disability ($\mu_{12}(t)$ in Examples 5, 8 and 9) can represent a lower risk of disablement thanks to a medical ascertainment carried out at policy issue. In any case, we can state that a duration-since-initiation dependence does not imply the use of more general and complicated models, since it is implicitly allowed by the Markov assumption for the process $\{S(t); t \geq 0\}$.

Problems might arise when considering model 2, the **duration-in-current state dependence**. It mainly concerns transitions from the disability state and requires **inception-select** transition intensities ('inception' denoting the time at which latest transition to that state occurred). Restricting our attention to transitions from the disability state, the intensities of recovery and mortality for disabled lives will be of interest. Of course, the Markovian property of the process $\{S(t); t \geq 0\}$ is lost. Nevertheless, the versatility of the Markov assumptions allows for practicable approaches to dealing with the duration-in-current-state dependence. The key idea is a redefinition of the state space, in such a way that states (or some states) also take into account information concerning the duration of presence. A general (and rather complicated) approach leads to the definition of semi-Markov processes (which we deal with briefly in this section), whilst a simpler (and more practicable) approach would require the 'splitting' of some states (see section 1.7).

Finally, the aim of model 3 is to stress the 'health story' of the insured. In general, taking into account aspects of the health story may lead to untractable models. However, some particular aspects can be introduced in actuarial models without dramatic consequences in terms of complexity. An interesting example will be presented and briefly discussed in section 1.7.

1.6.2 Definition of a semi-Markov process

Let us consider the stochastic process

$$\{S(t), R(t); t \geq 0\}$$

where $S(t)$ is the random state occupied by the risk at time t; $S(t)$ takes values in the state space $\mathscr{S}$, as defined in section 1.1; and $R(t)$ is the

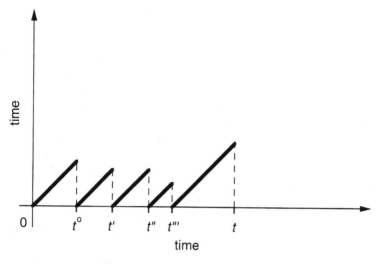

Fig. 1.14 A sample path of $\{R(t)\}$.

time spent in state $S(t)$ up to time t since the latest transition to that state; formally:

$$R(t) = \max\{\tau: \tau \leq t, S(t - h) = S(t) \text{ for all } h \in [0, \tau]\}.$$

$R(t)$ takes values in $[0 + \infty)$. Thus, the new stochastic process is defined by the pair of time-continuous stochastic processes $\{S(t)\}, \{R(t)\}$; it takes values in $\mathscr{S} \times [0, +\infty)$. Hence, $\mathscr{S} \times [0, +\infty)$ is the new state space.

A sample path of $\{R(t)\}$ is represented in Fig. 1.14; it corresponds to the sample path of $\{S(t)\}$ depicted in Fig. 1.13. Note that, for any given sequence of transition times $t^0, t', t'', \ldots$, the sample path of $\{R(t)\}$ is the same whatever the sequence of states visited by the process $\{S(t)\}$ may be.

Let us assume that $\{S(t), R(t); t \geq 0\}$ is a time-continuous, time-inhomogeneous Markov process. This means that, at time t, all conditional probabilities concerning the 'future' of the process after time t only depend on the 'most recent' information, i.e. the current state, $S(t) = i$, and the time elapsed since the latest transition into that state, $R(t) = r$. Thus, the aforesaid conditional probabilities do not depend on the information regarding the path of the process before time t, whatever this information may be. Hence, we use the following transition probabilities:

$$P_{ij}(t, u, r, w) = \Pr\{S(u) = j \wedge R(u) \leq w \,|\, S(t) = i \wedge R(t) = r\} \qquad (1.58)$$

for $0 \leq t < u$ and $r, w \geq 0$. Moreover, for $t = u$ we define

$$P_{ij}(t, u, r, w) = \delta_{ij}\varepsilon(w - r) \qquad (1.58')$$

where $\varepsilon(y) = 0$ or 1 according as $y < 0$ or $y \geq 0$. Then, the transition probabilities $P_{ij}(t, u, r, w)$ are defined for $0 \leq t \leq u$.

As $\{S(t), R(t); t \geq 0\}$ is a time-continuous, time-inhomogeneous Markov process, we say that $\{S(t); t \geq 0\}$ is a time-continuous, time-inhomogeneous semi-Markov process.

In order to interpret easily the assumption described above, consider the particular situation corresponding to the three-state model described in Example 9. Let i denote in equation (1.58) the state 'disabled', and j denote the state 'active'. According to our assumption, the probability that the insured is active at time u, with a time spent in the active state less than or equal to w, depends on the fact that at time t he is disabled with a time spent in disability equal to r, but takes no account of information such as whether he has experienced many (or few) lengthy (or short) periods of disability in the past.

Transition intensities are defined as follows:

$$\mu_{ij}(t, r) = \lim_{u \to t} \frac{P_{ij}(t, u, r, +\infty)}{u - t}. \tag{1.59}$$

We assume that the $\mu_{ij}(t, r)$'s exist for all $i \neq j$, $t \geq 0$, $r \geq 0$, and also that these intensities have properties which justify the operations we will perform.

Note that the expression $\mu_{ij}(t, r)\,dt$ can be interpreted as the probability that a transition from state i into state j occurs over the infinitesimal interval $[t, t + dt)$ given that the risk is in state i at time t with a time r spent in state i since the latest transition into that state.

In some cases, i.e. for some i, j, it may be reasonable to assume that the transition probability $P_{ij}(t, u, r, w)$ does not depend on time r, i.e.:

$$\Pr\{S(u) = j \wedge R(u) \leq w \mid S(t) = i \wedge R(t) = r\}$$
$$= \Pr\{S(u) = j \wedge R(u) \leq w \mid S(t) = i\}. \tag{1.60}$$

In these cases we simply write:

$$P_{ij}(t, u, w) = \Pr\{S(u) = j \wedge R(u) \leq w \mid S(t) = i\}. \tag{1.61}$$

It follows that the relevant intensities can be defined as:

$$\mu_{ij}(t) = \lim_{u \to t} \frac{P_{ij}(t, u, +\infty)}{u - t}. \tag{1.62}$$

Note that definition (1.62) exactly corresponds to definition (1.17) referring to the Markov process.

Moreover, we will use the following notation:

$$P_{ij}(t, u) = P_{ij}(t, u, +\infty) = \Pr\{S(u) = j \wedge R(u) < +\infty \mid S(t) = i\}. \tag{1.61'}$$

It is quite natural to define the occupancy probabilities within the semi-Markov framework. Indeed, we immediately find that:

$$\Pr\{S(z) = i \text{ for all } z \in [t, u] \mid S(t) = i \wedge R(t) = r\}$$
$$= \Pr\{S(u) = i \wedge R(u) = r + u - t \mid S(t) = i \wedge R(t) = r\}. \tag{1.63}$$

We will use the following notation:

$$P_{ii}(t, u, r) = \Pr\{S(u) = i \wedge R(u) = r + u - t \mid S(t) = i \wedge R(t) = r\}. \quad (1.64)$$

If, for a given state i, it is reasonable to assume that the occupancy probability $P_{ii}(t, u, r)$ does not depend on time r, then we will write:

$$P_{ii}(t, u) = \Pr\{S(u) = i \wedge R(u) \geq u - t \mid S(t) = i\}. \quad (1.65)$$

According to the TIA discussed in section 1.4.3, the transition probabilities must be derived, starting from the transition intensities. However, considerable complications arise when reasoning under semi-Markov assumptions. Therefore, we shall restrict our attention to a particular problem concerning disability insurance, in which the semi-Markov hypothesis really represents a step towards realistic modelling.

1.6.3 Example 9

Consider again the disability model, already discussed in sections 1.3 and 1.5. Assume that the transition intensities are as follows:

$\mu_{12}(t) = $ intensity of disability;

$\mu_{21}(t, r) = $ intensity of recovery;

$\mu_{13}(t) = $ intensity of mortality for active lives;

$\mu_{23}(t, r) = $ intensity of mortality for disabled lives.

Thus, transition intensities from the disability state are assumed to be inception-select, whilst transition intensities for an active life are assumed to be aggregate (or, possibly, issue-select; see section 1.6.1). Hence the stochastic process $\{S(t); t \geq 0\}$ is a time-continuous, time-inhomogeneous semi-Markov process. Note that the assumption concerning the intensities of transition from state 1 means that if the current state is 'active', then the future of the process is assumed to be independent of the duration of the current stay in that state.

Now, the simultaneous differential equations (1.36a) to (1.36f) (see section 1.5) must be replaced by a more complicated set of differential and integro-differential equations. For example, (1.36a) must be replaced by:

$$\frac{d}{dt}P_{11}(z, t) = \int_z^t P_{11}(z, v)\mu_{12}(v)P_{22}(v, t, 0)\mu_{21}(t, t - v)\,dv$$

$$- P_{11}(z, t)[\mu_{12}(t) + \mu_{13}(t)]. \quad (1.66)$$

The formal derivation of (1.66) is lengthy and beyond the scope of this section. To understand the structure of the formula at an intuitive level, note that the integral on the right-hand side of (1.66) replaces the term

$P_{12}(z,t)\mu_{21}(t)$ on the right-hand side of (1.36a). The integral is now needed since the transition intensity μ_{21} is a select intensity in the semi-Markov model. Hence the transitions $2 \to 1$ must be 'classified' according to the duration of the stay in state 2.

In addition, equation (1.37b) must be replaced by the following differential equation:

$$\frac{d}{dt}P_{\underline{22}}(z,t,r) = -P_{\underline{22}}(z,t,r)[\mu_{21}(t,t-z+r) + \mu_{23}(t,t-z+r)] \qquad (1.67)$$

whose solution is

$$P_{\underline{22}}(z,t,r) = \exp\left[-\int_z^t [\mu_{21}(v,v-z+r) + \mu_{23}(v,v-z+r)]\,dv\right]. \qquad (1.68)$$

1.7 SPLITTING OF STATES

1.7.1 Some preliminary ideas

The Markov model for $\{S(t); t \geq 0\}$ (see sections 1.4 and 1.5) involves consideration of the finite state space $\mathscr{S}$,

$$\mathscr{S} = \{1, 2, \ldots, N\}. \qquad (1.69)$$

The semi-Markov model is based on the (Markov) process $\{S(t), R(t); t \geq 0\}$, which involves consideration of the state space $\mathscr{S} \times [0, +\infty)$ instead of the finite state space $\mathscr{S}$. Note that we can write:

$$\mathscr{S} \times [0, +\infty) = \{1\} \times [0, +\infty) \cup \{2\} \times [0, +\infty) \cup \cdots \cup \{N\} \times [0, +\infty). \qquad (1.70)$$

In the TIA the probabilistic structure of a semi-Markov model is assigned in terms of the transition intensities $\mu_{ij}(t,r)$. However, the practical implementation of such a model actually requires select intensities $\mu_{ij}(t,r)$ just for some states i in $\mathscr{S}$, while for the remaining states in $\mathscr{S}$ the durational effect can be reasonably excluded and hence aggregate transition intensities (independent of r) can be assumed (see Example 9 in section 1.6).

In formal terms, the foundation of a model in which select and aggregate intensities coexist originates in a state space of the following type:

$$\mathscr{S}' = \{1\} \cup \{2\} \times [0, +\infty) \cup \cdots \cup \{N\}. \qquad (1.71)$$

It is self-evident that the state space $\mathscr{S}'$ represents an intermediate situation, between the two 'extreme' situations represented by (1.69) and (1.70). Example 9 in section 1.6, where

$$\mathscr{S}' = \{1\} \cup \{2\} \times [0, +\infty) \cup \{3\},$$

illustrates such an intermediate situation.

If we restrict the influence of the durational effect to some specified states, we can think of a different description of this effect. In particular, instead of reasoning in time-continuous terms (i.e. considering the interval $[0, +\infty)$ as the set of possible durations), we can think in time-discrete terms, thus considering a finite set of time labels $\{\tau_1, \tau_2, \ldots, \tau_m\}$, with $\tau_1 < \tau_2 < \cdots < \tau_m$.

In formal terms, the choice described above originates in a state space of the following type:

$$\mathscr{S}'' = \{1\} \cup \{2\} \times \{\tau_1, \tau_2, \ldots, \tau_m\} \cup \cdots \cup \{N\}. \tag{1.72}$$

A possible meaning of $\tau_1, \tau_2, \ldots, \tau_m$ (in terms of $\mathscr{S}''$) is as follows:

state $(2, \tau_1)$ = the risk is in state 2 of $\mathscr{S}$ with a duration of presence between 0 and τ_1;

state $(2, \tau_2)$ = the risk is in state 2 of $\mathscr{S}$ with a duration of presence between τ_1 and τ_2;

$$\vdots$$

state $(2, \tau_m)$ = the risk is in state 2 of $\mathscr{S}$ with a duration of presence greater than τ_{m-1}, where $\tau_m > \tau_{m-1}$.

According to this arrangement, for each j we should define m transition intensities:

$$\mu_{2j}(t; \tau_1), \mu_{2j}(t; \tau_2), \ldots, \mu_{2j}(t; \tau_m)$$

where $\tau_1, \tau_2, \ldots, \tau_m$ denote m duration parameters.

1.7.2 Back to the Markov model

The arrangement proposed above illustrates how to handle the duration of presence in time-discrete terms (the stochastic process $\{S(t); t \geq 0\}$ retaining its time-continuous set-up).

The same effect can be achieved as follows: Let us replace state 2 of $\mathscr{S}$ by the m states $2^{(1)}, 2^{(2)}, \ldots, 2^{(m)}$, where state $2^{(h)}$, $h = 1, 2, \ldots, m$, corresponds to state $(2, \tau_h)$ of the state space $\mathscr{S}''$ defined by equation (1.72). Thus, we have made a **splitting** of state 2 of the state space $\mathscr{S}$. The new state space is as follows:

$$\mathscr{S}^* = \{1, 2^{(1)}, 2^{(2)}, \ldots, 2^{(m)}, \ldots, N\}. \tag{1.73}$$

For each state j such that the transition from state 2 (in $\mathscr{S}$) into state j is possible, a set of m transition intensities must be assigned:

$$\mu_{2^{(1)}j}(t), \mu_{2^{(2)}j}(t), \ldots, \mu_{2^{(m)}j}(t).$$

The behaviour of these intensities will reflect the duration effect.

Of course, many transitions between states of $\mathscr{S}^*$ are impossible. For example, while $\mu_{12^{(1)}}(t) > 0$, the transition intensities $\mu_{12^{(h)}}(t)$, $h = 1, \ldots, m$, must be identically equal to zero.

Hence, we formally revert to the Markov model, in the sense that we again restrict our considerations to the Markov stochastic process $\{S(t); t \geq 0\}$ only, based on the state space $\mathscr{S}^*$. Thus, the introduction of more states representing the durational effect is a notational tool for treating semi-Markov models within the formally simpler Markov framework.

1.7.3 Example 9

Consider again the disability model, already discussed in sections 1.3, 1.5 and 1.6. Still restricting the selection effect to state 2 (i.e. the disability state), as in section 1.6, we now express this effect by splitting state 2 into m states (see Fig. 1.15), which represent disability according to duration since disablement. The meaning of the disability states is as follows:

$2^{(h)}$ = the insured is disabled with a duration of disability between $h - 1$ and h, for $h = 1, 2, \ldots, m - 1$;

$2^{(m)}$ = the insured is disabled with a duration of disability greater than $m - 1$.

The rationale of this splitting of the disability state is that, as is well known, for different durations since disability inception, observed

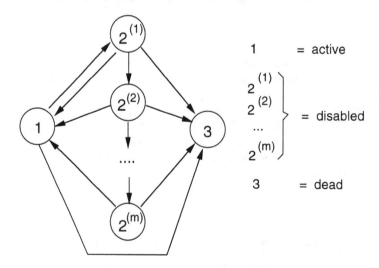

Fig. 1.15 A disability model with the disabled state split according to duration of disability.

recovery and mortality rates differ; i.e. the splitting allows us to consider select intensities (and probabilities) without formally introducing a semi-Markov model (leading to major difficulties).

The model requires the following transition intensities:

$$\mu_{13}(t), \mu_{12^{(1)}}(t)$$

$$\mu_{2^{(h)}1}(t), \mu_{2^{(h)}3}(t) \quad \text{for } h = 1, 2, \ldots, m.$$

As far as recovery is concerned, morbidity experience suggests:

$$\mu_{2^{(1)}1}(t) > \mu_{2^{(2)}1}(t) > \cdots > \mu_{2^{(m)}1}(t);$$

in particular, it is possible to put:

$$\mu_{2^{(h)}1}(t) > 0 \quad \text{for } h = 1, 2, \ldots, m - 1;$$

$$\mu_{2^{(m)}1}(t) = 0;$$

in this case, no recovery is possible after $m - 1$ years.

We do not consider here the problem of deriving transition probabilities from transition intensities. The reader should refer to section 1.8, in which this problem will be approached in more general terms.

The particular case in which $m = 2$ may be of special interest. Despite its simplicity, this four-state model can represent a realistic assumption. Think of the two disability states as follows:

$$\text{state } 2^{(1)} = \text{'unstable'};$$

$$\text{state } 2^{(2)} = \text{'stable'}.$$

In other words, the model allows for distinguishing between the 'acute' phase and the 'chronic' (albeit not necessarily permanent) phase of the disability spell. In the former, both recovery and mortality have high probabilities, while in the latter, recovery and mortality have lower probabilities. In terms of transition intensities, the following inequalities hold:

$$\mu_{2^{(1)}1}(t) > \mu_{2^{(2)}1}(t)$$

$$\mu_{2^{(1)}3}(t) > \mu_{2^{(2)}3}(t).$$

A further simplification can be obtained if we consider the 'stable' disability state as a state of attained permanent disability. In this case we assume

$$\mu_{2^{(2)}1}(t) = 0.$$

1.7.4 Example 9 (continuation)

We now illustrate a further possibility allowed by the multiple state modelling in a Markov framework, as far as disability annuities are concerned.

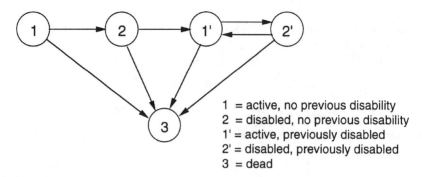

1 = active, no previous disability
2 = disabled, no previous disability
1' = active, previously disabled
2' = disabled, previously disabled
3 = dead

Fig. 1.16 A disability model with active and disabled states split according to previous disability.

Consider the five-state model depicted in Fig. 1.16. Also in this model, a splitting of states is performed. However, the splitting is no longer based on durations, but on the occurrence of (one or more) previous disability periods.

The rationale of splitting both the activity state and disability state is the assumption of a higher risk of disablement, a higher probability of death, and a lower probability of recovery as well, for an insured who has already experienced disability. In formal terms, the assumptions mentioned above can be expressed as follows:

disability: $\mu_{1'2'}(t) > \mu_{12}(t)$

recovery: $\mu_{2'1'}(t) < \mu_{21'}(t)$

death: $\mu_{1'3}(t) > \mu_{13}(t); \quad \mu_{2'3}(t) > \mu_{23}(t).$

1.8 FINDING TRANSITION PROBABILITIES

In Examples 1 to 11 of section 1.5 the derivation of transition probabilities from transition intensities (that is within the TIA), via the solution of a set of simultaneous differential equations, has been illustrated. The complexity of such a derivation is directly related to the particular structure of the underlying multiple state model, as clearly appears for instance comparing Examples 5, 8 and 9.

Of course, the problem of finding transition probabilities in the TIA must be approached in more general terms. In order to do this, we will approach this problem by following some procedures which are independent of the structure of the underlying multiple state model. However, it must be stressed that the procedures which we will consider require hypotheses concerning the functions used to represent the transition intensities. We will only give some basic ideas; the reader interested in more details should refer to the references quoted in the following.

1.8.1 Constant intensities (Jones, 1993)

We assume $\mu_{ij}(t) = \mu_{ij}$ for all t and for all $(i,j) \in \mathscr{T}$. This assumption leads to the well known time-homogeneous Markov process (see section 1.4), in which

$$\Pr\{S(u) = j \,|\, S(t) = i\} = P_{ij}(u - t).$$

Let us express the transition intensities and the transition probabilities in matrix form. Let

$$\mathbf{M} = ||\mu_{ij}||$$

where $\mathbf{M}$ denotes an $N \times N$ matrix with (i,j) entry μ_{ij} for $i \neq j$ and (i,i) entry $\mu_{ii} = -\mu_i$ (see section 1.4.3). Similarly, let

$$\mathbf{P}(z) = ||P_{ij}(z)|| \quad \text{for } z \geq 0$$

with

$$\mathbf{P}(0) = \mathbf{I}$$

where $\mathbf{I}$ denotes the identity matrix.

The Chapman–Kolmogorov equations may be written as follows:

$$\mathbf{P}(z) = \mathbf{P}(w)\mathbf{P}(z - w). \tag{1.74}$$

Denoting by $\mathbf{P}'(z)$ the matrix with (i,j) entry $(d/dz)P_{ij}(z)$, the Kolmogorov forward and backward differential equations may be respectively written as follows:

$$\mathbf{P}'(z) = \mathbf{P}(z)M \tag{1.75}$$

$$\mathbf{P}'(z) = -\mathbf{M}\mathbf{P}(z) \tag{1.76}$$

with boundary condition $\mathbf{P}(0) = \mathbf{I}$.

It is possible to prove that equations (1.75) and (1.76) have the following solution:

$$\mathbf{P}(z) = e^{\mathbf{M}z} = \mathbf{I} + \mathbf{M}z + \frac{\mathbf{M}^2 z^2}{2!} + \cdots. \tag{1.77}$$

However, the practical interest of equation (1.77) is rather limited, since the convergence of the series is slow. A more interesting procedure is suggested by Cox and Miller (1965). If $\mathbf{M}$ has distinct eigenvalues $d_1, \ldots, d_N$, then $\mathbf{M}$ can be expressed as follows:

$$\mathbf{M} = \mathbf{ADC} \tag{1.78}$$

where the ith column of $\mathbf{A}$, $i = 1, 2, \ldots, N$, is the eigenvector associated with d_i, $\mathbf{C} = \mathbf{A}^{-1}$, $\mathbf{D} = \text{diag}(d_1, \ldots, d_N)$. Furthermore:

$$\mathbf{P}(z) = \mathbf{A}\,\text{diag}(e^{d_1 z}, \ldots, e^{d_N z})\mathbf{C} \tag{1.79}$$

from which we finally get:

$$P_{ij}(z) = \sum_{h=1}^{N} a_{ih} c_{hj}\, e^{d_h z}.$$ (1.80)

Thus, the problem of finding the transition probability functions $P_{ij}(z)$ is reduced to the (numerical) problem of calculating the eigenvalues and eigenvectors of the transition intensity matrix $\mathbf{M}$.

1.8.2 Piecewise constant intensities (Jones, 1993)

As pointed out in section 1.4, in actuarial applications the seniority parameter is usually age-related, so that it is realistic to assume that the transition probabilities for any fixed period z vary in time. Hence, time-inhomogeneous models are required. We can accomplish this, while preserving the tractability of constant intensities, by using piecewise constant intensities.

First, for all i, j in $\mathscr{S}$, let $P_{ij}^{(m)}(z)$ denote the transition probability function associated with all time intervals contained in $(t_{m-1}, t_m]$, for $m = 1, 2, \ldots$, with $t_0 = 0$. Let us assume:

$$P_{ij}(t, t+z) = P_{ij}^{(m)}(z) \quad \text{if } t_{m-1} < t < t+z \le t_m.$$

For any t, let m_t denote the time interval which contains t.

Then, via the Chapman–Kolmogorov equations (see (1.12)) we can write:

$$P_{ij}(t, u) = \sum_{h_1 \in \mathscr{S}} \sum_{h_2 \in \mathscr{S}} \cdots \sum_{h_r \in \mathscr{S}} [P_{ih_1}^{(m_t)}(t_{m_t} - t)$$

$$\times P_{h_1 h_2}^{(m_t+1)}(t_{m_t+1} - t_{m_t}) \quad \cdots \quad P_{h_r j}^{(m_u)}(u - t_{m_u-1})] \quad (1.81)$$

where $r = m_u - m_t$. In matrix form, we can equivalently write:

$$\mathbf{P}(t, u) = \mathbf{P}^{(m_t)}(t_{m_t} - t) \; \mathbf{P}^{(m_t+1)}(t_{m_t+1} - t_{m_t}) \quad \cdots \quad \mathbf{P}^{(m_u)}(u - t_{m_u-1}).$$ (1.82)

Let $\mu_{ij}^{(m)}$ denote the (constant) transition intensity function pertaining to time interval m, i.e.:

$$\mu_{ij}(t) = \mu_{ij}^{(m)} \quad \text{if } t_{m-1} < t \le t_m.$$

Let, for $m = 1, 2, \ldots$:

$$\mathbf{M}^{(m)} = ||\mu_{ij}^{(m)}||$$

$$\mathbf{P}^{(m)}(z) = ||P_{ij}^{(m)}(z)||.$$

Then, following the procedure described in section 1.8.1, we can determine $\mathbf{A}^{(m)}$, $\mathbf{D}^{(m)}$, $\mathbf{C}^{(m)}$, and hence the transition probability matrix $\mathbf{P}^{(m)}(z)$ for any z, $z_{m-1} < z \leq z_m$, and $m = 1, 2, \ldots$:

$$\mathbf{P}^{(m)}(z) = \mathbf{A}^{(m)} \operatorname{diag}(e^{d_1^{(m)} z}, \ldots, e^{d_N^{(m)} z}) \mathbf{C}^{(m)}. \tag{1.83}$$

Moreover, relations (1.81) and (1.82) can be expressed in a simpler form, via the following recursion:

$$P_{ij}(t, u) = \sum_{h \in \mathscr{S}} P_{ih}(t, t_{m_u - 1}) P_{hj}^{(m_u)}(u - t_{m_u - 1}) \tag{1.84}$$

and also, in matrix form:

$$\mathbf{P}(t, u) = \mathbf{P}(t, t_{m_u - 1}) \mathbf{P}^{(m_u)}(u - t_{m_u - 1}). \tag{1.85}$$

Hence, letting

$$\mathbf{B}^{(m)} = \mathbf{P}(t, t_{m-1}) \mathbf{A}^{(m)}$$

we obtain

$$\mathbf{P}(t, u) = P(t, t_{m_u - 1}) \mathbf{A}^{(m_u)} \operatorname{diag}(e^{d_1^{(m_u)}(u - t_{m_u - 1})}, \ldots, e^{d_N^{(m_u)}(u - t_{m_u - 1})}) \mathbf{C}^{(m_u)}$$

$$= \mathbf{B}^{(m_u)} \operatorname{diag}(e^{d_1^{(m_u)}(u - t_{m_u - 1})}, \ldots, e^{d_N^{(m_u)}(u - t_{m_u - 1})}) \mathbf{C}^{(m_u)}$$

and finally:

$$P_{ij}(t, u) = \sum_{h=1}^{N} b_{ih}^{(m_u)} c_{hj}^{(m_u)} e^{d_h^{(m_u)}(u - t_{m_u - 1})}. \tag{1.86}$$

1.8.3 Continuous intensities over intervals (Wolthuis, 1994)

We now express the transition probabilities $P_{ij}(t, u)$ in terms of the transition intensities, assuming that the intensities are continuous functions over $[t, u]$. We only give a sketch of the procedure; the reader interested in a deeper presentation can refer to Wolthuis (1994).

Let

$$P_{ij}^{[k]}(t, u) = \Pr\{S(u) = j \text{ after exactly k transitions in } [t, u] \mid S(t) = i\}. \tag{1.87}$$

For $k = 0$ we have:

$$P_{ij}^{[0]}(t, u) = \begin{cases} 0 & \text{if } i \neq j \\ P_{jj}(t, u) & \text{if } i = j \end{cases}$$

i.e.

$$P_{ij}^{[0]}(t, u) = \delta_{ij} \exp\left[-\int_t^u \mu_j(z) \, dz\right]. \tag{1.88}$$

For $k \geq 1$ the following recursions hold:

$$P_{ij}^{[k]}(t, u) = \sum_{h:\, h \neq j} \int_t^u P_{ih}^{[k-1]}(t, z) \mu_{hj}(z) P_{\underline{jj}}(z, u) \, dz \tag{1.89}$$

$$P_{ij}^{[k]}(t, u) = \sum_{h:\, h \neq j} \int_t^u P_{\underline{ii}}(t, z) \mu_{ih}(z) P_{hj}^{[k-1]}(z, u) \, dz. \tag{1.90}$$

Recursions (1.89) and (1.90) can be easily interpreted by direct reasoning. From (1.89) we immediately obtain, for $k = 1$:

$$P_{ij}^{[1]}(t, u) = \int_t^u P_{\underline{ii}}(t, z) \mu_{ij}(z) P_{\underline{jj}}(z, u) \, dz \tag{1.91}$$

and for $k = 2$:

$$\begin{aligned} P_{ij}^{[2]}(t, u) &= \sum_{h:\, h \neq j} \int_t^u P_{ih}^{[1]}(t, z) \mu_{hj}(z) P_{\underline{jj}}(z, u) \, dz \\ &= \sum_{h:\, h \neq j} \int_t^u \int_t^z P_{\underline{ii}}(t, w) \mu_{ih}(w) P_{\underline{hh}}(w, z) \mu_{hj}(z) P_{\underline{jj}}(z, u) \, dw \, dz \end{aligned} \tag{1.92}$$

(from which a set of 'repeated integration formulae' follows). It should be stressed that the procedure allows us to resort to the calculation of probabilities of the following type only (i.e. occupancy probabilities):

$$P_{\underline{gg}}(v, w) = \exp\left[-\int_v^w \mu_g(z) \, dz \right].$$

It is possible to prove that the transition probabilities can be expressed as an infinite sum of probabilities $P^{[k]}$; formally:

$$P_{ij}(t, u) = \sum_{k=0}^{+\infty} P_{ij}^{[k]}(t, u). \tag{1.93}$$

In particular, it can be proved that probabilities $P_{ij}(t, u)$ as defined by equation (1.93) satisfy the Chapman–Kolmogorov equations if the transition intensities are continuous functions over $[t, u]$.

Now, let us assume that the transition intensities are continuous over each time interval $[t_0, t_1), [t_1, t_2), \ldots$, with $t_0 = 0$. The procedure sketched above allows us to determine probabilities $P^{[k]}$ and then the matrix **P** relating to these intervals and to any related subinterval.

In order to calculate the transition probabilities $P_{ij}(t, u)$ when $[t, u]$ is not included in any interval $[t_s, t_{s+1})$, we can use the Chapman–Kolmogorov equations to express $P_{ij}(t, u)$ in terms of the already determined transition probabilities. First, let us define for $s = 1, 2, \ldots$:

$$P_{ij}(t, t_s) = \lim_{u \to t_s} P_{ij}(t, u); \quad t_{s-1} \leq t < t_s. \tag{1.94}$$

Now, we can use a relation quite similar to equation (1.81) (see section 1.8.2):

$$P_{ij}(t,u) = \sum_{h_1 \in \mathscr{S}} \sum_{h_2 \in \mathscr{S}} \cdots \sum_{h_r \in \mathscr{S}} [P_{ih_1}^{(m_t)}(t, t_{m_t})$$

$$\times P_{h_1 h_2}^{(m_t + 1)}(t_{m_t}, t_{m_t + 1}) \cdots P_{h_r j}^{(m_u)}(t_{m_u - 1}, u)] \qquad (1.95)$$

where m_t denotes the time interval which contains t, $r = m_u - m_t$ and

$$P_{ij}^{(m)}(v,w) = P_{ij}(v,w); \quad t_{m-1} \leq v < w \leq t_m.$$

1.8.4 Example 9

Let us consider again the disability model, allowing for not necessarily permanent disability, with state space $\mathscr{S} = \{1,2,3\}$ (see Fig. 1.9). Let us assume that the transition intensities are continuous over each interval $[m, m+1)$, $m = 0,1,\ldots$, i.e. over each policy year. Then we can use the procedure described in section 1.8.3, in order to find the transition probabilities. Moreover, let us restrict ourselves to a maximum of three transitions a year. For any $[t, u]$ included in $[m, m+1)$, we then find:

$$P_{12}(t,u) = P_{12}^{[1]}(t,u) + P_{12}^{[3]}(t,u)$$

$$P_{11}(t,u) = P_{11}^{[0]}(t,u) + P_{11}^{[2]}(t,u) = P_{11}(t,u) + P_{11}^{[2]}(t,u)$$

$$P_{13}(t,u) = 1 - P_{11}(t,u) - P_{12}(t,u)$$

$$\vdots$$

(note that, of course, $P_{12}^{[2]}(t,u) = 0$, $P_{11}^{[1]}(t,u) = 0, \ldots,$). For example, we have (see equation (1.92)):

$$P_{11}^{[2]}(t,u) = \int_t^u \int_t^z P_{11}(t,w)\mu_{12}(w)P_{22}(w,z)\mu_{21}(z)P_{11}(z,u)\,dw\,dz.$$

1.9 INCREMENT-DECREMENT TABLES

Just as in life insurance mathematics, where the life table is a convenient method for summarizing the underlying transition probabilities (e.g. in the case of Example 1), so for multiple state models a convenient method of presentation is the increment–decrement table (or **multiple state life table**). We shall present a brief discussion here of this concept in terms of Example 9, and the corresponding three-state model.

The fundamental quantities in the model are the transition intensities. Let us assume that these have been estimated from an investigation. Then, as discussed earlier, we can use the transition intensities to calculate the transition probabilities and can use these transition probabilities to construct an increment–decrement table based on this model.

To do this, let x_0 be an appropriate initial age. Let $l_x^{(1)}$ denote the expected number of lives in state 1 at age x and $l_x^{(2)}$ the expected number of lives in state 2 at age x for $x \geq x_0$. Then $l_{x_0}^{(1)}$ and $l_{x_0}^{(2)}$ are arbitrary and determine an initial distribution for the underlying stochastic process in the sense that

$$\frac{l_{x_0}^{(i)}}{l_{x_0}^{(1)} + l_{x_0}^{(2)}} \quad \text{for } i = 1, 2$$

may be regarded as the probability that a life aged x_0 is in state i. We note that this is in direct analogy with the life table where the radix l_α is arbitrary so that the life table forms a relative scale determined by this initial l_α value.

The increment–decrement table numbers of lives in each state are specified by $kl_x^{(1)}$ and $kl_x^{(2)}$ where k is an arbitrary constant and

$$l_x^{(1)} = \frac{l_{x_0}^{(1)} P_{11}(x_0, x) + l_{x_0}^{(2)} P_{21}(x_0, x)}{l_{x_0}^{(1)} + l_{x_0}^{(2)}} \tag{1.96'}$$

$$l_x^{(2)} = \frac{l_{x_0}^{(1)} P_{12}(x_0, x) + l_{x_0}^{(2)} P_{22}(x_0, x)}{l_{x_0}^{(1)} + l_{x_0}^{(2)}}. \tag{1.96''}$$

We now need to consider how $l_x^{(1)}$ and $l_x^{(2)}$ change as we move from age x to age $x + 1$. We shall let ${}^i d_x^{jk}$ be the expected number of moves from state j to state k in the interval $(x, x + 1)$, among the population consisting of the closed group of those in state i at x (numbering $l_x^{(1)}$ for $i = 1$ and $l_x^{(2)}$ for $i = 2$). Then we shall consider the expected number of transitions that occur in the year of age $(x, x + 1)$. We shall let $\mu_{jk}(0)$ refer to the initial age x_0.

The expected number of transitions between states 1 and 2 (i.e. incidences of disability) will be

$$^1d_x^{12} + {}^2d_x^{12} = l_x^{(1)} \int_0^1 P_{11}(x, x + t)\mu_{12}(x - x_0 + t)\,dt$$

$$+ l_x^{(2)} \int_0^1 P_{21}(x, x + t)\mu_{12}(x - x_0 + t)\,dt. \tag{1.97a}$$

The expected number of transitions between states 2 and 1 (i.e. recoveries from the disabled state) will be

$$^1d_x^{21} + {}^2d_x^{21} = l_x^{(1)} \int_0^1 P_{12}(x, x + t)\mu_{21}(x - x_0 + t)\,dt$$

$$+ l_x^{(2)} \int_0^1 P_{22}(x, x + t)\mu_{21}(x - x_0 + t)\,dt. \tag{1.97b}$$

The expected number of transitions between states 1 and 3 (i.e. deaths while healthy) will be

$$^1d_x^{13} + {}^2d_x^{13} = l_x^{(1)} \int_0^1 P_{11}(x, x + t)\mu_{13}(x - x_0 + t)\,dt$$

$$+ l_x^{(2)} \int_0^1 P_{21}(x, x + t)\mu_{13}(x - x_0 + t)\,dt. \qquad (1.97c)$$

The expected number of transitions between states 2 and 3 (i.e. deaths while disabled) will be

$$^1d_x^{23} + {}^2d_x^{23} = l_x^{(1)} \int_0^1 P_{12}(x, x + t)\mu_{23}(x - x_0 + t)\,dt$$

$$+ l_x^{(2)} \int_0^1 P_{22}(x, x + t)\mu_{23}(x - x_0 + t)\,dt. \qquad (1.97d)$$

Given values of the transition intensities from an investigation, it would be possible to calculate numerically each of the integrals and hence estimate the $^i d_x^{jk}$ functions. The numbers of transitions calculated via the above equations could then be used for entries in an increment–decrement life table. Thus, it is possible to demonstrate that

$$l_{x+1}^{(1)} = l_x^{(1)} + ({}^1d_x^{21} + {}^2d_x^{21}) - ({}^1d_x^{12} + {}^2d_x^{12}) - ({}^1d_x^{13} + {}^2d_x^{13}) \qquad (1.98')$$

$$l_{x+1}^{(2)} = l_x^{(2)} + ({}^1d_x^{12} + {}^2d_x^{12}) - ({}^1d_x^{21} + {}^2d_x^{21}) - ({}^1d_x^{23} + {}^2d_x^{23}). \qquad (1.98'')$$

Equations (1.98) would then be consistent with the definitions in equations (1.96). Clearly equations (1.98) are generalizations of the simple recurrence relation that holds for the life table, namely

$$l_{x+1} = l_x - d_x$$

and they can be written in words as

Healthy at $x + 1 =$ Healthy at $x +$ Recoveries $-$ Incidences
$-$ Deaths as healthy

Disabled at $x + 1 =$ Disabled at $x +$ Incidences $-$ Recoveries
$-$ Deaths as disabled.

1.10 ACTUARIAL VALUES OF BENEFITS

In section 1.2 benefits and premiums have been introduced. Several examples have been presented in section 1.3. As further steps in describing the insurance contract, we need to:

1. assume applicable 'principles' in order to determine premiums and reserves for any given set of benefits;
2. give formulae for the calculation of premiums;

3. give formulae for the calculation of reserves.

In order to perform these steps, the concept of 'expected present value', briefly 'actuarial value', must be defined. This concept is dealt with in the present section, while section 1.11 deals with points 1, 2 and 3.

1.10.1 Random present values

First, we have to define random present values. As far as the financial structure is concerned, we assume the compound interest model with a deterministic, constant force of interest δ. We denote by v, $v = e^{-\delta}$, the annual discount factor.

Let I_E denote the indicator of the event E, i.e.:

$$I_E = \begin{cases} 1 & \text{if } E \text{ is true} \\ 0 & \text{if } E \text{ is false.} \end{cases} \tag{1.99}$$

We now define the random present values of benefits and premiums which have been described in section 1.2.

1. Consider a continuous annuity benefit at a rate $b_j(u)$ at time u, if $S(u) = j$; thus, $b_j(u)\,du$ is the benefit amount paid out in the infinitesimal interval $[u, u + du)$. Then, the random present value at time t is:

$$Y_t(u, u + du) = v^{u-t}I_{\{S(u)=j\}}b_j(u)\,du. \tag{1.100}$$

Moreover, the random present value of the annuity benefit on the time interval $[u_1, u_2)$, with $t \le u_1 < u_2$, is:

$$Y_t(u_1, u_2) = \int_{u_1}^{u_2} v^{u-t}I_{\{S(u)=j\}}b_j(u)\,du. \tag{1.101}$$

2. Consider a lump sum $c_{jk}(u)$ paid just after time u if a transition $j \to k$ occurs at time u; let $\{S(u^-) = j \wedge S(u) = k\}$ denote the transition. Hence, for any given u the random present value at time t is:

$$Y_t(u) = v^{u-t}I_{\{S(u^-)=j\wedge S(u)=k\}}c_{jk}(u). \tag{1.102}$$

Now consider the time interval $(u_1, u_2]$ and assume that a lump sum $c_{jk}(u)$ is paid at each transition $j \to k$ in $(u_1, u_2]$. Let $N_{jk}(u)$ denote the random number of transitions $j \to k$ in $(0, u]$. The random present value of the transition benefits is given by:

$$Y_t(u_1, u_2) = \int_{u_1}^{u_2} v^{u-t}c_{jk}(u)\,dN_{jk}(u). \tag{1.103}$$

3. Consider a lump sum benefit $d_j(u)$ ('pure endowment') at some fixed time u if $S(u) = j$. The random present value at time t is:

$$Y_t(u) = v^{u-t}I_{\{S(u)=j\}}d_j(u). \tag{1.104}$$

4. As far as random present values of continuous premiums and discrete-time premiums are concerned, we can immediately refer to cases 1 and 3 presented above, simply replacing $b_j(u), d_j(u)$ by $p_j(u), \pi_j(u)$ respectively.

1.10.2 Actuarial values

Actuarial values are expected present values and, as noted in section 1.10.1, are required for the determination of premiums and reserves. In order to determine expected values, we also need a probabilistic structure (which is not required while defining random present values). To this purpose, we assume that the stochastic process $\{S(t); t \geq 0\}$ (the state space being $\mathscr{S}$ or $\mathscr{S}^*$) is a time-continuous Markov chain.

Now let us suppose that at time t, $t \geq 0$, the risk is in the state i, i.e. $S(t) = i$. We define the actuarial values of benefits and premiums considered above, assuming $S(t) = i$ as a conditioning event. Hence, actuarial values are conditional expectations.

1. Actuarial values are given by:

$$E[Y_t(u, u + du) \mid S(t) = i] = v^{u-t} P_{ij}(t, u) b_j(u) \, du \tag{1.105}$$

$$E[Y_t(u_1, u_2) \mid S(t) = i] = \int_{u_1}^{u_2} v^{u-t} P_{ij}(t, u) b_j(u) \, du. \tag{1.106}$$

In particular, for a unit-level annuity benefit paid in state j during the period $[t, n)$, the following notation is used:

$$\bar{a}_{ij}(t, n) = \int_t^n v^{u-t} P_{ij}(t, u) \, du. \tag{1.107}$$

2. Actuarial values are given by:

$$E[Y_t(u) \mid S(t) = i] = v^{u-t} P_{ij}(t, u) \mu_{jk}(u) c_{jk}(u) \, du \tag{1.108}$$

$$E[Y_t(u_1, u_2) \mid S(t) = i] = \int_{u_1}^{u_2} v^{u-t} P_{ij}(t, u) \mu_{jk}(u) c_{jk}(u) \, du. \tag{1.109}$$

For $c_{jk}(u) = 1$, the following notations are used:

$$\bar{A}_{ijk}(t, n) = \int_t^n P_{ij}(t, u) \mu_{jk}(u) v^{u-t} \, du \tag{1.110}$$

$$\bar{A}_{i.k}(t, n) = \sum_{j:\, j \neq k} \bar{A}_{ijk}(t, n) \tag{1.111}$$

$$\bar{A}_{ij.}(t, n) = \sum_{k:\, k \neq j} \bar{A}_{ijk}(t, n). \tag{1.112}$$

3. The actuarial value is:

$$E[Y_t(u) \mid S(t) = i] = v^{u-t} P_{ij}(t, u) d_j(u). \tag{1.113}$$

For $d_j(u) = 1$, the following notation is used:

$$\bar{E}_{ij}(t,u) = v^{u-t} P_{ij}(t,u). \tag{1.114}$$

Moreover, if $c_{jk}(u) = 1$ for all u in $(t,n]$ and $d_j(m) = 1$, the notation is as follows:

$$\bar{A}_{ij.}(t,n;m) = \bar{A}_{ij.}(t,n) + \bar{E}_{ij}(t,m). \tag{1.115}$$

When $m = n$ we have a case of special interest, generalizing the well-known endowment insurance.

4. In order to determine the actuarial values of continuous and single premiums, simply use equations (1.105), (1.106) and (1.113), replacing $b_j(u)$ by $p_j(u)$ and $d_j(u)$ by $\pi_j(u)$ respectively.

1.10.3 An example of approximate evaluation

First, let us consider the actuarial value given by formula (1.107). Assume that t and n are integer times (i.e. policy anniversaries). We can write:

$$\bar{a}_{ij}(t,n) = \int_t^n v^{u-t} P_{ij}(t,u)\, du \cong \sum_{h=t+1}^n \left[v^{h-1/2-t} \int_{h-1}^h P_{ij}(t,u)\, du \right]. \tag{1.116}$$

Note that the integral in brackets represents the conditional expected value at time t of the random time spent in state j during the policy year $(h-1,h)$. Denote this by $\bar{e}_{ij}(t;h-1,h)$. Then:

$$\bar{a}_{ij}(t,n) \cong \sum_{h=t+1}^n v^{h-1/2-t}\, \bar{e}_{ij}(t;h-1,h). \tag{1.117}$$

Now, let us consider a three-state model, with $1 =$ active, $2 =$ disabled, $3 =$ dead. Hence, $\bar{a}_{12}(t,n)$ refers to a disability annuity. Note that $P_{12}(t,n)$, which appears in equation (1.116), can be expressed as follows:

$$P_{12}(t,u) = P_{11}(t,h-1)P_{12}(h-1,u) + P_{12}(t,h-1)P_{22}(h-1,u).$$

Then we have:

$$\begin{aligned}
\bar{a}_{ij}(t,n) \cong & \sum_{h=t+1}^n \Big[v^{h-1/2-t} \Big(P_{11}(t,h-1) \int_{h-1}^h P_{12}(h-1,u)\, du \\
& + P_{12}(t,h-1) \int_{h-1}^h P_{22}(h-1,u)\, du \Big) \Big] \\
= & \sum_{h=t+1}^n [v^{h-1/2-t}(P_{11}(t,h-1)\bar{e}_{12}(h-1;h-1,h) \\
& + P_{12}(t,h-1)\bar{e}_{22}(h-1;h-1,h))].
\end{aligned}$$

Finally, assume

$$\bar{e}_{12}(h-1;h-1,h) = \bar{e}_{22}(h-1;h-1,h) \tag{1.119}$$

and denote this value by $\bar{e}_2(h-1,h)$; thus, $\bar{e}_2(h-1,h)$ represents the expected time spent in sickness between $h-1$ and h by an insured who is alive at time $h-1$ (disregarding the state then occupied). We obtain:

$$\bar{a}_{ij}(t,n) \cong \sum_{h=t+1}^{n} [v^{h-1/2-t}(P_{11}(t,h-1) + P_{12}(t,h-1))\bar{e}_2(h-1,h)]. \tag{1.120}$$

Formulae like this are sometimes used in actuarial practice in order to calculate expected present values of disability benefits. Their use can be justified when statistical data are scanty so that transition intensities cannot be estimated, whilst data like $\bar{e}_2(h-1,h)$ (sometimes called **persistency rates**) are available. Moreover, when this approximation is used, the quantity $(P_{11}(t,h-1) + P_{12}(t,h-1))$, which represents the probability for a life active at time t of being alive at time $h-1$, is usually replaced by a simple survival probability, so disregarding the state at time t (however, such a probability is not well-defined within the context of a Markov model).

Remark
We shall return to other examples of approximate evaluation that are used practically in Chapter 3 when we discuss in more depth the three state 'active-disabled-dead' model.

1.10.4 An interesting property

Let us consider the actuarial value, at time t, of a deferred level annuity benefit of one unit paid in state j during the period $[\tau, n]$, with $\tau > t$. This actuarial value is given by (see equation (1.106)):

$$\int_{\tau}^{n} v^{u-t} P_{ij}(t,u) \, du = \sum_{h \in \mathscr{S}} v^{\tau-t} P_{ih}(t,\tau) \int_{\tau}^{n} v^{u-\tau} P_{hj}(\tau,u) \, du$$

$$= \sum_{h \in \mathscr{S}} \bar{E}_{ih}(t,\tau)\bar{a}_{hj}(\tau,n). \tag{1.121}$$

Thus, we can express the actuarial value of a deferred level annuity benefit in terms of the actuarial values of non-deferred level annuity benefits. Note that the property presented above follows from:

- the Chapman–Kolmogorov relation;
- the compound interest property, i.e. $v^z = v^u v^{z-u}$.

The property can be used in order to derive formulae regarding other types of actuarial values, For instance, we have

$$
\begin{aligned}
\bar{A}_{ijk}(t,n) &= \int_t^n P_{ij}(t,u)\mu_{jk}(u)v^{u-t}\,du \\
&= \int_t^\tau P_{ij}(t,u)\mu_{jk}(u)v^{u-t}\,du + \int_\tau^n P_{ij}(t,u)\mu_{jk}(u)v^{u-t}\,du \\
&= \bar{A}_{ijk}(t,\tau) + \int_\tau^n \sum_{h\in\mathscr{S}} P_{ih}(t,\tau)P_{hj}(\tau,u)\mu_{jk}(u)v^{u-t}\,du \\
&= \bar{A}_{ijk}(t,\tau) + \sum_{h\in\mathscr{S}} P_{ih}(t,\tau)v^{\tau-t}\int_\tau^n P_{hj}(\tau,u)\mu_{jk}(u)v^{u-\tau}\,du \\
&= \bar{A}_{ijk}(t,\tau) + \sum_{h\in\mathscr{S}} \bar{E}_{ih}(t,\tau)\bar{A}_{hjk}(\tau,n).
\end{aligned} \tag{1.122}
$$

1.11 PREMIUMS AND RESERVES

Let us refer to an insurance policy providing all the benefits defined in section 1.2. Let n denote the terms of the policy. The actuarial value at time t, if $S(t) = i$, of these benefits is:

$$
\begin{aligned}
\mathscr{B}_i(t,n) &= \int_t^n v^{u-t}\left[\sum_{j\in\mathscr{S}} P_{ij}(t,u)b_j(u)\right]du \\
&+ \int_t^n v^{u-t}\left[\sum_{j\in\mathscr{S}}\sum_{k:\,k\neq j} P_{ij}(t,u)\mu_{jk}(u)c_{jk}(u)\right]du \\
&+ \sum_{u:\,u\geq t} v^{u-t}\left[\sum_{j\in\mathscr{S}} P_{ij}(t,u)d_j(u)\right].
\end{aligned} \tag{1.123}
$$

The actuarial value at time t, if $S(t) = i$, of the premiums paid on a continuous basis is:

$$
\mathscr{P}_i(t,n) = \int_t^n v^{u-t}\left[\sum_{j\in\mathscr{S}} P_{ij}(t,u)p_j(u)\right]du \tag{1.124}
$$

while the actuarial value of discrete-time premiums is:

$$
\Pi_i(t,n) = \sum_{u:\,u\geq t} v^{u-t}\left[\sum_{j\in\mathscr{S}} P_{ij}(t,u)\pi_j(u)\right]. \tag{1.125}
$$

As a premium calculation principle, we assume the **equivalence principle**. By definition, the equivalence principle is fulfilled if and only if:

$$
\mathscr{P}_1(0,n) = \mathscr{B}_1(0,n) \tag{1.126}
$$

or

$$\Pi_1(0, n) = \mathscr{B}_1(0, n). \tag{1.127}$$

It must be stressed that, for any given set of benefits, conditions (1.126) and (1.127) are generally satisfied by an infinity of premium functions. Hence, further conditions are required in order to assess premiums (for example: premiums increasing or decreasing according to a given law, level premiums, etc.). For the discussion of some examples, refer to section 1.12.

Furthermore, it must be pointed out that the insurer should be 'funded' by each insurance contract, in the sense that premium timing should determine, at any time, a debt position for the insurer himself. Formally, for any t:

$$\mathscr{B}_{S(t)}(t, n) \geq \mathscr{P}_{S(t)}(t, n). \tag{1.128}$$

Then, the 'funding condition' represents a constraint on premium assessment.

The prospective reserve at time t is defined as the actuarial value of future benefits less the actuarial value of future premiums. Formally, if $S(t) = i$, the prospective reserve $\bar{V}_i(t)$ is given by:

$$\bar{V}_i(t) = \mathscr{B}_i(t, n) - \mathscr{P}_i(t, n) \tag{1.129}$$

or by:

$$\bar{V}_i(t) = \mathscr{B}_i(t, n) - \Pi_i(t, n). \tag{1.130}$$

Note that a specific reserve is defined for each state i which may be occupied by the risk at time t.

In terms of the prospective reserve, as $S(0) = 1$, the equivalence principle is satisfied if and only if $\bar{V}_1(0) = 0$. The funding requirement can be expressed as:

$$\bar{V}_{S(t)}(t) \geq 0. \tag{1.131}$$

For the sake of simplicity, let us now refer to an insurance policy providing benefits of the types $b_j(u)$ and $c_{jk}(u)$ only, with continuous premiums $p_j(u)$. Then, the prospective reserve is given by:

$$\bar{V}_i(t) = \int_t^n v^{u-t} \left[\sum_{j \in \mathscr{S}} P_{ij}(t, u) b_j(u) \, du + \sum_{j \in \mathscr{S}} \sum_{k: k \neq j} P_{ij}(t, u) \mu_{jk}(u) c_{jk}(u) du \right.$$

$$\left. - \sum_{j \in \mathscr{S}} P_{ij}(t, u) p_j(u) \, du \right]. \tag{1.132}$$

Differentiation of equation (1.132) with respect to t ($0 < t < n$) gives the following differential equation:

$$\frac{d}{dt} \bar{V}_i(t) = \delta \bar{V}_i(t) - b_i(t) + p_i(t) - \sum_{j: j \neq i} \mu_{ij}(t)[c_{ij}(t) + \bar{V}_j(t) - \bar{V}_i(t)]. \tag{1.133}$$

This equation generalizes **Thiele's differential equation**, discovered in 1875 for the case of a single-life assurance. Note that equation (1.133) holds for all $i \in \mathscr{S}$, namely a set of simultaneous differential equations is involved. The interpretation of (1.133) is easier when the equation is written in differential form. Let us assume that in state i the continuous benefit is equal to zero, i.e. $b_i(t) = 0$. Then, we have:

$$d\bar{V}_i(t) = \bar{V}_i(t)\delta \, dt + p_i(t) \, dt - \sum_{j:j\neq i} [c_{ij}(t) + \bar{V}_j(t) - \bar{V}_i(t)]\mu_{ij}(t) \, dt. \quad (1.134)$$

The right-hand side of equation (1.134) exhibits positive and negative contributions determining the increment of the reserve, i.e. $d\bar{V}_i(t)$. Positive contributions come from interest, $\bar{V}_i(t)\delta \, dt$, and from premium payment, $p_i(t) \, dt$; the third contribution, due to transition payments, may be positive or negative, depending on $c_{ij}(t)$ (which may be equal to zero) as well as on the sign of the reserve jump, $\bar{V}_j(t) - \bar{V}_i(t)$. The term in brackets is usually called the **sum at risk**.

Finally, equation (1.134) can be written as follows:

$$p_i(t) \, dt = [d\bar{V}_i(t) - \bar{V}_i(t)\delta \, dt] + \sum_{j:j\neq i} [c_{ij}(t) + \bar{V}_j(t) - \bar{V}_i(t)]\mu_{ij}(t) \, dt. \quad (1.135)$$

Hence, a decomposition of the premium paid in $[t, t + dt)$ is obtained. The first term on the right-hand side is called the **savings premium**; it represents the increment in the reserve, excluding the effect of additions through interest. The second term is called the **risk premium** and represents the expected cost related to the sum at risk.

1.12 EXAMPLES

This section is concerned with premium and reserve calculations. The examples that we discuss are drawn from the set of examples presented in section 1.3, and in section 1.5 as regards the probabilistic structures.

1.12.1 Example 1

Consider a temporary assurance with a constant sum assured c; let n denote the term of the policy. Assume a continuous premium at an instantaneous rate $p_1(u)$ (of course $p_1(u) = 0$ for $u > n$). The equivalence principle requires that the premium function satisfies the following equation:

$$\int_0^n v^u P_{11}(0, u) p_1(u) \, du = c \int_0^n v^u P_{11}(0, u) \mu_{12}(u) \, du \quad (1.136)$$

i.e.

$$\int_0^n v^u P_{11}(0, u) p_1(u) \, du = c\bar{A}_{112}(0, n). \quad (1.136')$$

In order to determine a particular solution, we must restrict ourselves to some class of functions $p_1(u)$. For example, we can choose $p_1(u) = p$ for $0 \leq u < n$. Then, we find:

$$p = c \frac{\bar{A}_{112}(0, n)}{\bar{a}_{11}(0, n)}. \tag{1.137}$$

Now, let us assume a single premium (at time 0). The solution is trivial:

$$\pi_1(0) = c\bar{A}_{112}(0, n). \tag{1.138}$$

Finally, let us consider a discrete-time premium arrangement, assuming that premiums are paid at policy anniversaries. Thus, we have to choose a sequence $\pi_1(0), \pi_1(1), \ldots, \pi_1(n-1)$ satisfying the following equation:

$$\sum_{h=0}^{n-1} v^h P_{11}(0, h)\pi_1(h) = c\bar{A}_{112}(0, n). \tag{1.139}$$

It can be easily proved that a particular solution of equation (1.139) is given by:

$$\pi_1(h) = c\bar{A}_{112}(h, h+1); \quad h = 0, 1, \ldots, n-1. \tag{1.140}$$

It is self-evident that each of the premiums defined by equation (1.140) exactly covers the annual expected cost of the insurer, valued at time h. In this case, the $\pi_1(h)$s are often called **natural premiums**.

As far as the prospective reserve is concerned, first note that the reserve itself is only defined for $S(t) = 1$, since the policy ceases just after entering state 2. In the case of a continuous premium, we have:

$$\bar{V}_1(t) = c\bar{A}_{112}(t, n) - \int_t^n v^u P_{11}(t, u)p_1(u)\, du \tag{1.141}$$

and, if $p_1(u) = p$:

$$\bar{V}_1(t) = c\bar{A}_{112}(t, n) - p\bar{a}_{11}(t, n). \tag{1.142}$$

If a single premium is paid, we trivially have, for $0 < t \leq n$:

$$\bar{V}_1(t) = c\bar{A}_{112}(t, n). \tag{1.143}$$

When single discrete-time premiums are concerned, each premium being given by formula (1.140), we obviously find:

$$\bar{V}_1(h) = 0; \quad h = 0, 1, \ldots, n \tag{1.144}$$

while for $h < t < h+1, h = 0, 1, \ldots, n-1$, we have:

$$\bar{V}_1(t) = c\bar{A}_{112}(t, h+1). \tag{1.144'}$$

1.12.2 Example 3

Consider a deferred continuous annuity at a rate b after time m and until the death of the insured. Let us assume a continuous premium at an instantaneous rate $p_1(u)$ (of course $p_1(u) = 0$ for $u > m$). The equivalence principle requires that the premium function satisfies the following equation:

$$\int_0^m v^u P_{11}(0, u) p_1(u)\, du = b \int_m^{+\infty} v^u P_{11}(0, u)\, du, \qquad (1.145)$$

i.e. (see equation (1.121)):

$$\int_0^m v^u P_{11}(0, u) p_1(u)\, du = b \bar{E}_{11}(0, m) \bar{a}_{11}(m, +\infty). \qquad (1.145')$$

A particular solution of equation (1.145') is obtained setting $p_1(u) = p$; we find:

$$p = b \frac{\bar{E}_{11}(0, m) \bar{a}_{11}(m, +\infty)}{\bar{a}_{11}(0, m)}. \qquad (1.146)$$

Now, let us consider a discrete-time premium arrangement, assuming that premiums are paid at policy anniversaries. Premiums $\pi_1(0)$, $\pi_1(1), \ldots, \pi_1(m-1)$ must satisfy the following equation:

$$\sum_{h=0}^{m-1} v^h P_{11}(0, h) \pi_1(h) = b \bar{E}_{11}(0, m) \bar{a}_{11}(m, +\infty). \qquad (1.147)$$

Note that any particular sequence of premiums can be interpreted as follows. By means of premium $\pi_1(h)$, $h = 0, 1, \ldots, m-1$, a part of the insured benefit b, say $b^{(h)}$, is purchased. Thus, $b^{(h)}$ is the benefit accrued in year $(h, h+1)$, and the following relation holds:

$$\pi_1(h) = b^{(h)} \bar{E}_{11}(h, m) \bar{a}_{11}(m, +\infty); \quad h = 0, 1, \ldots, m-1. \qquad (1.148)$$

Of course, $\sum_{h=0}^{m-1} b^{(h)} = b$, and the $\pi_1(h)$s satisfy equation (1.147).

The prospective reserve, in the case of a continuous premium, is given by:

$$\bar{V}_1(t) = \begin{cases} b \bar{E}_{11}(t, m) \bar{a}_{11}(m, +\infty) - \int_t^m v^u P_{11}(t, u) p_1(u)\, du, & \text{if } t < m \\ b \bar{a}_{11}(t, +\infty), & \text{if } t \geq m \end{cases}$$

$$(1.149)$$

and, if $p_1(u) = p$:

$$\bar{V}_1(t) = \begin{cases} b \bar{E}_{11}(t, m) \bar{a}_{11}(m, +\infty) - p \bar{a}_{11}(t, m), & \text{if } t < m \\ b \bar{a}_{11}(t, +\infty), & \text{if } t \geq m. \end{cases} \qquad (1.150)$$

In the case of discrete-time premiums, from the general definition (1.130) we obtain, for $t < m$:

$$\bar{V}_1(t) = b_1 v^{m-t} P_{11}(t, m) \bar{a}_{11}(m, +\infty) - \sum_{u=\lceil t \rceil}^{m-1} v^{u-t} P_{11}(t, u) \pi_1(u) \qquad (1.151)$$

where $\lceil t \rceil$ denotes the smallest integer greater than or equal to t. After some manipulations, we have:

$$\bar{V}_1(t) = \sum_{h=0}^{\lceil t \rceil - 1} b^{(h)} \bar{E}_{11}(t, m) \bar{a}_{11}(m, +\infty) \quad \text{if } t < m \qquad (1.151')$$

(for $t \geq m$ see (1.149) and (1.150)). Note that formula (1.151') can be immediately explained by direct reasoning according to the accrued benefit scheme, as for equations (1.140) and (1.147) earlier. Formula (1.151') must actually be used when the sequence of premiums is not completely planned at policy issue.

1.12.3 Example 9

Consider an annuity benefit in the case of total disability. Let n denote the policy term. Let the stopping time (from policy issue) of the annuity payment be equal to n. Let b denote the constant rate of the disability annuity. Assume a time-continuous premium, of course payable for n years at most. The equivalence principle is fulfilled if the premium function $p_1(u)$ satisfies the following equation:

$$\int_0^n v^u P_{11}(0, u) p_1(u) \, du = b \int_0^n v^u P_{12}(0, u) \, du \qquad (1.152)$$

i.e.

$$\int_0^n v^u P_{11}(0, u) p_1(u) \, du = b \bar{a}_{12}(0, n). \qquad (1.152')$$

An example of a premium function is given by:

$$p_1(u) = \begin{cases} p & \text{if } 0 \leq t < m \\ 0 & \text{if } t \geq m \end{cases} \qquad (1.153)$$

where $m \leq n$. In this case equation (1.152') can be expressed as follows:

$$p \bar{a}_{11}(0, m) = b \bar{a}_{12}(0, n). \qquad (1.152'')$$

In the case of a single premium, we simply have:

$$\pi_1(0) = b \bar{a}_{12}(0, n). \qquad (1.154)$$

Note that the right-hand side member of equation (1.152) expresses the actuarial value of the insured benefit in terms of the probability of being disabled at time u. From equation (1.40), we obtain:

$$P_{12}(0, u) = \int_0^u P_{11}(0, s)\mu_{12}(s)P_{\underline{22}}(s, u)\, ds \qquad (1.155)$$

whence

$$\int_0^n v^u P_{12}(0, u)\, du = \int_0^n \int_0^u v^u P_{11}(0, s)\mu_{12}(s)P_{\underline{22}}(s, u)\, ds\, du \qquad (1.156)$$

and finally, changing the order of integration:

$$\int_0^n v^u P_{12}(0, u)\, du = \int_0^n P_{11}(0, s)\mu_{12}(s)v^s \left[\int_s^n P_{\underline{22}}(s, u)v^{u-s}\, du \right] ds. \qquad (1.157)$$

The right-hand side member of equation (1.157) expresses the actuarial value of the insured benefit in terms of the probability of becoming disabled at time s (given by $\mu_{12}(s)\, ds$) and the probability of remaining disabled from time s to time u (given by $P_{22}(s, u)$). Moreover, note that the integral in brackets represents the actuarial value (at time u) of a (temporary) disability annuity paid until death or recovery. As we will see in Chapter 3, several calculation methods used in the actuarial practice of disability insurance can be interpreted in the light of formulae (1.152) and (1.157).

As regards the prospective reserve, two reserves must be defined, pertaining to the 'active' state (state 1) and the 'disabled' state (state 2) respectively. In the case of a continuous premium, we have:

$$\bar{V}_1(t) = b \int_t^n v^{u-t}P_{12}(t, u)\, du - \int_t^n v^{u-t}P_{11}(t, u)p_1(u)\, du$$

$$= b\bar{a}_{12}(t, n) - \int_t^n v^{u-t}P_{11}(t, u)p_1(u)\, du \qquad (1.158)$$

$$\bar{V}_2(t) = b \int_t^n v^{u-t}P_{22}(t, u)\, du - \int_t^n v^{u-t}P_{21}(t, u)p_1(u)\, du$$

$$= b\bar{a}_{22}(t, n) - \int_t^n v^{u-t}P_{21}(t, u)p_1(u)\, du. \qquad (1.159)$$

In the case of a single premium, we simply have (for $t > 0$):

$$\bar{V}_1(t) = b\bar{a}_{12}(t, n) \qquad (1.158')$$

$$\bar{V}_2(t) = b\bar{a}_{22}(t, n). \qquad (1.159')$$

We now present a numerical example, in which the calculations have been performed assuming constant transition intensities. The relevant formulae used to derive the transition probabilities from the transition intensities can be found in section 1.5.7.

Let $x = 40$ be the age at entry. Let $n = 10$ be the policy term. The constant rate of the disability annuity is $b = 1$. Let us assume that a

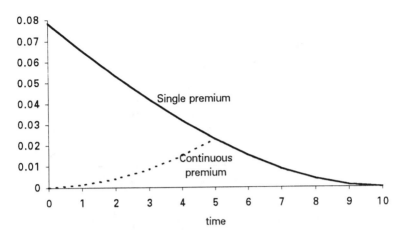

Fig. 1.17 Active reserve; $x = 40$, $n = 10$, $m = 5$.

continuous constant premium p is paid for $m = 5$ years while the insured is active; hence the premium function given by (1.153) is used. Let $\delta = \ln 1.04$ be the force of interest. Assume the following transition intensities:

$$\mu_{12} = 0.002136; \quad \mu_{13} = 0.004183; \quad \mu_{21} = 0.005; \quad \mu_{23} = 1.2\mu_{13} = 0.005020.$$

The values of μ_{12} and μ_{13} have been calculated by averaging the values of the corresponding intensities in the Danish disability model (see section 1.5.5). The value of μ_{21} and the extra level of mortality for disabled lives are simple assumptions, because the Danish model does not allow for recovery and does not consider extra mortality.

The resulting actuarial value of the benefits, i.e. the single premium $\pi_1(0)$, is equal to 0.07839. Then we find $p = 0.01753$. Figure 1.17 illustrates the behaviour of the active reserve, $\bar{V}_1(t)$, in the case of a single premium and of a continuous constant premium as well. Note that the two reserves coincide for $t \geq m$. In Fig. 1.18 the behaviour of the disabled reserve, $\bar{V}_2(t)$, for the case of a single premium is depicted.

Now let us consider the following example: $x = 40$, $n = 15$, $m = 18$. The premium function given by (1.153) is still used. In order to take into account the different age interval now involved, the transition intensities must be changed. To this purpose, let us assume the following transition intensities:

$$\mu_{12} = 0.003287; \quad \mu_{13} = 0.005285; \quad \mu_{21} = 0.005; \quad \mu_{23} = 1.2\mu_{13} = 0.006343.$$

The values of μ_{12}, μ_{13} (and hence μ_{23}) have still been calculated averaging the corresponding intensities in the Danish model, now considering the ages 40 to 55.

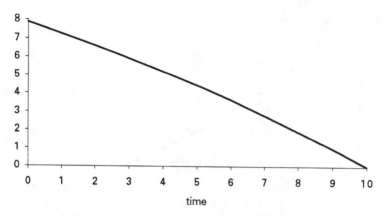

Fig. 1.18 Disabled reserve (single premium); $x = 10$, $n = 10$.

The resulting actuarial value of the benefits, i.e. the single premium $\pi_1(0)$, is equal to 0.2297. Then we find $p = 0.03455$. Figures 1.19 and 1.20 illustrate the behaviour of the active reserve and of the disabled reserve respectively.

1.12.4 Example 10a

Following on from section 1.12.3 we consider an annuity benefit in the case of unemployment. The policy term is n years but the benefit, paid continuously at a constant rate b, is paid for a maximum period of one year measured from the time of entry to state 2. The single premium (at time 0) is then given by

$$\pi_1(0) = b \int_0^n v^s P_{11}(0, s) \mu_{12}(s) \left[\int_s^{s+1} P_{\underline{22}}(s, u) v^{u-s} \, du \right] ds. \qquad (1.160)$$

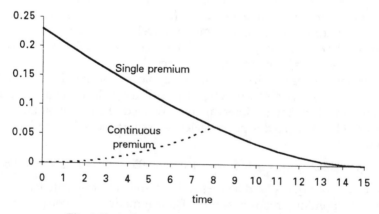

Fig. 1.19 Active reserve; $x = 40$, $n = 15$, $m = 8$.

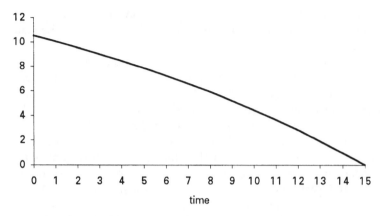

Fig. 1.20 Disabled reserve (single premium); $x = 40$, $n = 15$.

As an alternative, we could consider an annuity benefit of b payable continuously during the first spell of unemployment, if any, that the policyholder suffers. Entitlement to benefit ceases at time n, i.e. the stopping time (from policy issue) of the annuity payment is equal to n. Let $\bar{a}_{t|}$ denote the present value of an annuity certain of 1 per annum, payable for t years:

$$\bar{a}_{t|} = \int_0^t v^s \, ds = \frac{1 - e^{-\delta t}}{\delta}. \tag{1.161}$$

The single premium (at time 0) is then given by

$$\pi_1(0) = b \int_0^n v^s P_{\underline{11}}(0, s)\mu_{12}(s)$$

$$\times \left[\int_s^n \bar{a}_{u-s|} P_{\underline{22}}(s, u)\mu_{21}(u) \, du + \bar{a}_{n-s|} P_{\underline{22}}(s, n) \right] ds \tag{1.162}$$

which can be shown to be equivalent to

$$\pi_1(0) = b \int_0^n v^s P_{\underline{11}}(0, s)\mu_{12}(s) \left[\int_s^n P_{\underline{22}}(s, u)v^{u-s} \, du \right] ds. \tag{1.163}$$

1.13 DISTRIBUTIONS OF RANDOM PRESENT VALUES

1.13.1 Some preliminary ideas

Although actuarial calculations for life and disability covers traditionally deal with expected values only (such as premiums determined according to the equivalence principle and reserves), a more complete stochastic analysis can be performed (at least in principle), based on the probabilistic model which describes the path of the insured risk.

In particular, knowledge of the distribution of the present value of benefits less premiums can give insight into the riskiness of the insurance contract.

In the following chapters we will restrict our attention to calculations of expected present values according to the traditional actuarial approach, albeit in the context of multiple state modelling. Nevertheless, in this section we provide a brief outline about the construction of the probability distribution of the present value of benefits less premiums.

Probability distributions related to standard single-life insurance benefits, depending only on the residual lifetime (pure endowment, temporary assurance, endowment assurance, life annuity), have been studied by De Pril (1989). Distributions of present values related to a 'general life insurance', defined as a combination of the most common life benefits, have been analysed by Dhaene (1990). Other authors deal with the computation of moments; see for example the authors cited by Dhaene (1990).

It should be stressed that in these cases the underlying probabilistic structure is simply given by the two-state model used in Examples 1 to 3 (see sections 1.3, 1.5 and 1.12). Of course, the construction of the probability distribution is much more complicated when a general multiple state model is concerned. For a general approach to the calculation of distributions of present values in a multiple state framework, the reader is referred to Hesselager and Norberg (1996).

Some insight into the riskiness of the insurance contract can be provided also by the consideration of higher moments of the random present value, e.g. variance and skewness. Of course, a simpler methodology can be used when only the calculation of moments is involved.

In this section we simply illustrate the construction of a probability distribution related to a very simple three-state model. To this purpose, we consider the disability policy providing a lump sum benefit in the case of permanent and total disability (see Example 5 in sections 1.3 and 1.5).

1.13.2 Example 5

Consider an n-year insurance policy providing a lump sum benefit in the case of permanent and total disability. Let c denote the sum assured. Premiums are assumed to be paid continuously at the rate of p over $[0, n)$ when the contract stays in state 1. The equivalence principle leads to:

$$p\bar{a}_{11}(0, n) = c\bar{A}_{112}(0, n). \tag{1.164}$$

Let L denote the random present value (at time 0) of benefits less premiums; the quantity L is usually called the 'loss function'. Of course (according to the notation defined in section 1.11),

$$E(L) = \mathcal{B}_1(0, n) - \mathcal{P}_1(0, n) \tag{1.165}$$

so that the equivalence principle is fulfilled if and only if the expected value of the loss function is equal to 0.

Let T denote the random time at which the risk leaves state 1, entering state 2 or state 3. Let $\bar{a}_{t\rceil}$ denote the present value of an annuity certain of 1 per annum, payable for t years (see equation (1.161)). We find:

(S1)   if $\{T = 0$ with transition $1 \to 2\}$, then $L = c$;
(S2)   if $\{T = 0$ with transition $1 \to 3\}$, then $L = 0$;
(S3)   if $\{T = t, 0 < t \le n$, with transition $1 \to 2\}$, then $L = cv^t - p\bar{a}_{t\rceil}$;
(S4)   if $\{T = t, 0 < t \le n$, with transition $1 \to 3\}$, then $L = -p\bar{a}_{t\rceil}$;
(S5)   if $\{t > n\}$, then $L = -p\bar{a}_{n\rceil}$.

Events involved in statements (S1) to (S5) constitute a partition of the outcome space described in terms of the possible values of T and possible transitions from state 1.

The following inequalities hold for $0 < t < n$:

$$-p\bar{a}_{n\rceil} < -p\bar{a}_{t\rceil} < 0 < cv^t - p\bar{a}_{t\rceil} < cv^n - p\bar{a}_{n\rceil} < c \qquad (1.166)$$

(in particular, note that $cv^n - q\bar{a}_{n\rceil} = 0$ where q is the instantaneous rate of payment funding the deterministic lump sum c at time n, namely $q > p$).

Let $t_1(z)$ denote the time such that the loss function L takes the value $z = -p\bar{a}_{t_1(z)\rceil}$; from equation (1.161) we find:

$$t_1(z) = \frac{1}{\delta}\log\frac{p}{p + \delta z}.$$

Let $t_2(z)$ denote the time such that the loss function L takes the value $z = cv^t - p\bar{a}_{t_2(z)\rceil}$; we obtain:

$$t_2(z) = \frac{1}{\delta}\log\frac{p + c\delta}{p + \delta z}.$$

Let

$$\Phi_L(z) = \Pr\{L \le z\} \qquad (1.167)$$

denote the probability distribution function of the random variable L. By direct reasoning, we find:

$$\Phi_l(z) = \begin{cases} 0 & \text{if } z < -p\bar{a}_{n\rceil} \\ P_{11}(0, n) + P_{11}(0, t_1(z))P_{13}(t_1(z), n) & \text{if } -p\bar{a}_{n\rceil} \le z < 0 \\ P_{11}(0, n) + P_{13}(0, n) & \text{if } 0 \le z < cv^n - p\bar{a}_{n\rceil} \\ \begin{aligned} &P_{11}(0, n) + P_{13}(0, n) \\ &\quad + P_{11}(0, t_2(z))P_{12}(t_2(z), n) \end{aligned} & \text{if } cv^n - p\bar{a}_{n\rceil} \le z < c \\ 1 & \text{if } z \ge c. \end{cases}$$

$$(1.168)$$

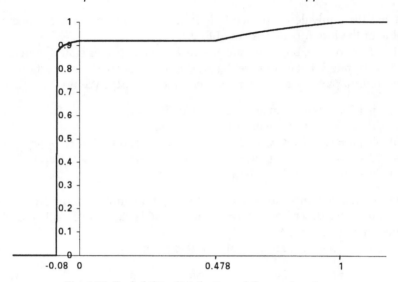

Fig. 1.21 Probability distribution of the random loss.

Let us consider the following numerical example. Let $x = 40$ be the age at entry; let $n = 15$ be the policy term. The lump sum is $c = 1$. Let $\delta = \ln 1.04$ be the force of interest. Assume the following transition intensities (see section 1.5.5):

$$\mu_{12}(t) = 0.0004 + 10^{0.06(x+t)-5.46}$$

$$\mu_{13}(t) = 0.0005 + 10^{0.038(x+t)-4.12}.$$

We find $p = 0.007037$. Figure 1.21 illustrates the behaviour of the function $\Phi_L(z)$. The jump at $-p\bar{a}_{n\rceil} = -0.08$ is due to the positive probability of being alive and active at age $x + n = 55$. Moreover, note that $0.478 = cv^n - p\bar{a}_{n\rceil}$ and, obviously, $1 = c$.

Some comments can help us to understand how the probability distribution of the random loss can be used in performing actuarial calculations. First, note that the distribution illustrated by Fig. 1.21 follows from the use of the equivalence principle in the premium calculation. Of course, different distributions can be constructed starting from different premium levels. In particular, the upper tail of the distribution corresponding to different premiums can be analysed. Conversely, the probability distribution Φ_L of the random loss can be used as a tool to implement different premium principles. For example, according to the percentile principle, leading to the so-called 'chance constrained premium' (see, for example, Gerber, 1979), the premium $p^*(\varepsilon)$ is defined as follows:

$$p^*(\varepsilon) = \min\{p \mid \Phi_L(0) \geq 1 - \varepsilon\} \tag{1.169}$$

where ε is a given 'small' probability. In words: $p^*(\varepsilon)$ is the smallest premium such that the probability of a (positive) loss is at most ε.

1.14 REFERENCES AND SUGGESTIONS FOR FURTHER READING

In this section we only quote papers and textbooks dealing with general aspects of multiple state modelling in life insurance and related fields. Studies particularly devoted to applications of multiple state modelling to disability annuities, long-term care insurance, critical illness cover, etc., will be quoted in the relevant sections of the following chapters.

The first studies of the Markov approach to life insurance actuarial problems seem to be due to Franckx (1963), Daboni (1964) and Amsler (1968).

The seminal paper of Hoem (1969a) places life and other contingencies within the framework of a general, unified, probabilistic theory, using the Markov assumption. A time-continuous approach is adopted and formulae and theorems for actuarial values, premiums and reserves are derived.

Recently, multiple state models have been used as a powerful tool for dealing with many actuarial problems in life insurance and related fields. Very interesting examples are provided by Amsler (1988), Hoem (1988), Norberg (1988, 1991), Ramlau-Hansen (1988, 1991), Waters (1984) and Wilkie (1998b). Applications of multiple state modelling are also discussed by Gatenby and Ward (1994) and Jones (1993, 1994).

Stochastic rates of interest in the framework of a multiple state Markov model are studied by Norberg (1995).

An interesting application of multiple state Markov modelling to non-life insurance problems is proposed by Hesselager (1994). Hoem (1977) deals with working life tables within a Markov context.

The first application of semi-Markov models to disability insurance seems to be due to Janssen (1966). The use of semi-Markov processes in actuarial science and in demography is discussed by Hoem (1972).

The first book dealing with the insurances of the person within a multiple state framework is probably due to De Vylder (1973). A modern textbook on life insurance mathematics, in which a continuous time Markov model is adopted, is due to Wolthuis (1994). Readers are referred to this textbook also for selected references.

For further discussion of multiple decrement models and their application to pension funds, readers are referred to Neill (1977) and Bowers *et al.* (1986).

Multistate life table methodology is discussed from a probabilistic point of view by Hoem and Funck Jensen (1982).

For further discussion of increment–decrement tables, readers should consult Haberman (1983, 1984).

2

Multiple state models for life and other contingencies: the time-discrete approach

---

2.1 THE TIME-DISCRETE MARKOV MODEL

2.1.1 Some preliminary ideas

Let us assume that a time-continuous Markov model has been assigned. Thus, the transition intensities have been specified and the transition probabilities have been derived. It is self-evident that the implied time-discrete probabilistic structure can be immediately derived. To do this, we simply have to restrict our attention to transition probabilities $P_{ij}(z, t)$ only, where z, t now denote integer values (actually, the term 'time-discrete' typically refers to a scanning of time on an annual or monthly basis).

It must be stressed that, conversely, the construction of a time-continuous Markov model given a time-discrete one is not possible, at least without assuming appropriate hypotheses. For example, consider a three-state model, $\mathscr{S} = \{1, 2, 3\}$, with $1 = $ 'active', $2 = $ 'disabled', $3 = $ 'dead'. How would we calculate

$$\Pr\{(S(z + 1) = 3) \wedge (S(t) = 2 \text{ for some } t \in (z, z + 1)) \mid S(z) = 1\}?$$

To assess the above probability, for example the following set of hypotheses might be assumed:

- no more than two transitions within the unit period, i.e. the year;
- uniform distribution of first transition time within the year;
- the probability that the second transition occurs within the second half of the year is equal to one half of the probability that a transition of the same type occurs within the year.

The hypotheses mentioned above, or others like these, are often adopted in disability actuarial models.

Of course, a time-discrete Markov model has practical interest apart from its possible derivation from a time-continuous one. For this reason, we will now give an autonomous definition of the time-discrete Markov model. Later (and in particular in Chapter 3, section 3.3, where some examples concerning disability annuities shall be presented) we will discuss how to deal with time-continuous actuarial problems using a time-discrete model as a starting point. The importance of this is due to the fact that, in actuarial practice, the choice of a discrete framework does not correspond to the assumption that transitions can occur at integer times only, but simply expresses a yearly based scanning of time.

2.1.2 Definition of a time-discrete Markov chain

Consider a time-discrete stochastic process $\{S(t); t = 0, 1, 2, \ldots\}$, with a finite state space $\mathscr{S}$. We say that $\{S(t); t = 0, 1, 2, \ldots\}$ is a **time-discrete Markov chain** if, for any n and each finite set of integer times $(0 \leq) t_0 < \cdots < t_{n-1} < t_n < u$ and corresponding set of states $i_0, \ldots i_{n-1}$, i_n, j in $\mathscr{S}$ with

$$\Pr\{S(t_0) = i_0 \wedge \cdots \wedge S(t_{n-1}) = i_{n-1} \wedge S(t_n) = i_n \wedge S(u) = j\} > 0 \quad (2.1)$$

the following property (the so-called Markov property) is satisfied:

$$\Pr\{S(u) = j \,|\, S(t_0) = i_0 \wedge \cdots \wedge S(t_{n-1}) = i_{n-1} \wedge S(t_n) = i_n\}$$
$$= \Pr\{S(u) = j \,|\, S(t_n) = i_n\}. \quad (2.2)$$

Thus, as for the time-continuous case, it is assumed that the conditional probability at the left-hand side of equation (2.2) only depends on the 'most recent' information $S(t_n) = i_n$ and is independent of the path before t_n.

The conditional probabilities $\Pr\{S(u) = j \,|\, S(t) = i\}$ are called transition probabilities, also in a time-discrete context. They are usually denoted by:

$$P_{ij}(t, u) = \Pr\{S(u) = j \,|\, S(t) = i\}. \quad (2.3)$$

These probabilities satisfy the probability conditions (see equations (1.10a) and (1.10b)) and the **Chapman–Kolmogorov equations**:

$$P_{ij}(t, u) = \sum_{k \in \mathscr{S}} P_{ik}(t, w) P_{kj}(w, u) \quad (2.4)$$

where t, w, u are now integer times ($t \leq w \leq u$).

Thanks to the Chapman–Kolmogorov equations, the following recursion in particular holds:

$$P_{ij}(t, u) = \sum_{k \in \mathscr{S}} P_{ik}(t, t+1) P_{kj}(t+1, u). \quad (2.5)$$

Hence, all probabilities $P_{ij}(t, u)$ can be derived from the set of one-year transition probabilities $P_{ij}(t, t+1)$. These probabilities can be denoted as follows:

$$P_{ij}[z] = P_{ij}(z, z+1); \quad z = 0, 1, 2, \ldots \tag{2.6}$$

In particular, when a time-homogeneous process is concerned, we simply have:

$$P_{ij}[z] = P_{ij}; \quad z = 0, 1, 2, \ldots \tag{2.7}$$

We shall denote with $P_{\underline{ii}}(t, u)$ the probability of remaining in state i up to time u, given that the risk is in state i at time t. If we assume (as is rather usual in actuarial practice) that no more than one transition can occur during one year (apart from the possible death of the insured), we simply have:

$$P_{\underline{ii}}(t, u) = \prod_{h=t}^{u-1} P_{ii}(h, h+1) = \prod_{h=t}^{u-1} P_{ii}[h] \tag{2.8}$$

2.2 EXAMPLES

The examples that we discuss in this section are drawn from the set of examples presented in Chapter 1. The grouping reflects the structure of the graph which describes states and transitions.

2.2.1 Examples 1, 2, 3

In these examples the randomness only comes from the lifetime of the insured. The following probabilities must be assigned, for $t = 0, 1, 2, \ldots$:

$$P_{11}[t], \quad P_{12}[t], \quad \text{with } P_{11}[t] + P_{12}[t] = 1.$$

Note that, according to the traditional actuarial notation, we have:

$$P_{11}[t] = \begin{cases} {}_1p_{x+t} & \text{if an aggregate model is used} \\ {}_1p_{[x]+t} & \text{if an issue-select model is used} \end{cases}$$

$$P_{12}[t] = \begin{cases} {}_1q_{x+t} & \text{if an aggregate model is used} \\ {}_1q_{[x]+t} & \text{if an issue-select model is used} \end{cases}$$

where x denotes the (integer) age at policy issue.

Remark
For the 'alive–dead' model, it is rather easy to build a time-continuous structure given a time-discrete one, when proper assumptions are made. Let y denote an integer age, and let $0 < r < 1$. Consider the following three assumptions, which are well known and widely used in traditional life insurance mathematics.

1. Assume:

$$_rq_y = r \cdot {}_1q_y \qquad (2.9)$$

This assumption is known as the **uniform distribution of deaths** throughout the year of age.
2. Assume:

$$_{1-r}q_{y+r} = (1-r) \, {}_1q_y \qquad (2.10)$$

Formula (2.10) is called the **Balducci hypothesis**.
3. A third mortality assumption is that of a **constant intensity of mortality** over a given unit age interval:

$$\mu_{y+r} = \mu. \qquad (2.11)$$

In this case we have:

$$_{1-r}q_{y+r} = 1 - e^{-\mu(1-r)}. \qquad (2.12)$$

2.2.2 Examples 4, 5

In these examples a three-state model is concerned, with a strictly transient state (state 1) and two absorbing states (states 2 and 3). The following probabilities describe the stochastic structure of the model:

$$P_{11}[t], \quad P_{12}[t], \quad P_{13}[t] \quad \text{with } P_{11}[t] + P_{12}[t] + P_{13}[t] = 1.$$

In Example 4 (temporary assurance with a rider benefit in case of accidental death), $P_{13}[t]$ is the probability of accidental death, while $P_{12}[t]$ is the probability of death due to other causes. $P_{11}[t]$ is, of course, the survival probability.

In Example 5 (lump sum benefit in case of permanent and total disability), $P_{11}[t]$ is the probability of remaining an active or non-permanently and totally disabled life. As far as the meaning of $P_{13}[t]$ is concerned, it is very important to stress that this quantity is not the probability of dying between ages $x + t$ and $x + t + 1$. Actually, it is the probability of leaving the state 1 because of death. Note that death might occur between ages $x + t$ and $x + t + 1$, after disablement; of course the sequence disablement–death is not 'included' in $P_{13}[t]$. Similarly, $P_{12}[t]$ must be interpreted as the probability of leaving the state 1 because of disablement. Note that the events 'transition out of state 1 because of death' and 'transition out of state 1 because of disablement' are mutually exclusive events, and hence the relevant probabilities can be added, the sum $P_{12}[t] + P_{13}[t]$ representing the probability of leaving state 1 between ages $x + t$ and $x + t + 1$.

It should be stressed that similar distinctions are meaningless within a time-continuous context. In particular, since no more than one transition can occur in the infinitesimal interval $[t, t + dt)$, the quantity $\mu_{13} \, dt$ actually represents the probability of dying between ages $x + t$ and $x + t + dt$.

2.2.3 Example 8

This example concerns an annuity benefit in case of permanent and total disability. Hence it is based on a three-state model, with two strictly transient states (states 1 and 2) and an absorbing state (state 3). Still $P_{11}[t]$ is the probability of remaining an active or non-permanently and totally disabled life (if we assume that no more than one transition can occur in a one-year period, apart from death). Now, $P_{13}[t]$ is the probability of being dead at time $t + 1$, and hence the probability of dying between ages $x + t$ and $x + t + 1$ disregarding the state (active or disabled) occupied at time of death. Conversely, $P_{12}[t]$ is the probability of being disabled at time $t + 1$. Note that it cannot be interpreted as the probability of disablement between ages $x + t$ and $x + t + 1$, since disablement can be followed by death within the year. Finally, $P_{23}[t]$ is the probability of dying between ages $x + t$ and $x + t + 1$ for a disabled person at age $x + t$.

2.2.4 Example 9

This example is based on a three-state model, with two transient states (states 1 and 2) and an absorbing state (state 3). If we assume a homogeneous process, i.e. with $P_{ij}(z, z + 1) = P_{ij}[z] = P_{ij}$, then the matrix of one-year transition probabilities, which has the form

$$P = \begin{pmatrix} P_{11} & P_{12} & P_{13} \\ P_{21} & P_{22} & P_{23} \\ 0 & 0 & 1 \end{pmatrix} \tag{2.13}$$

completely defines the stochastic behaviour of the risk.

Let $P_{ij}^{(h)}$ denote the (i, j) entry of the power matrix P^h (which is the identity matrix if $h = 0$). Then, if we assume the homogeneity hypothesis we simply have

$$P_{ij}(0, h) = P_{ij}^{(h)}.$$

Of course the homogeneity hypothesis is unrealistic when insurances of the person are dealt with, as transition probabilities do depend on the attained age of the insured. To express this dependence, we have to work with a time-inhomogeneous Markov process, whose probabilistic structure is defined by the sequence of matrices of one-year transition probabilities. A matrix of one-year transition probabilities has the form

$$P[t] = \begin{pmatrix} P_{11}[t] & P_{12}[t] & P_{13}[t] \\ P_{21}[t] & P_{22}[t] & P_{23}[t] \\ 0 & 0 & 1 \end{pmatrix}; \quad t = 0, 1, \ldots \tag{2.14}$$

Furthermore, if we assume the (realistic) hypothesis that different durations since disability inception imply different probabilities of recovery and death, then we have to resort to more complicated models. An example will be presented in the next section.

2.3 SPLITTING OF STATES

Let us consider a disability model (see Fig. 1.15) in which the state 2, i.e. the disability state, is replaced by a set of disability states $2^{(1)}, 2^{(2)}, \ldots, 2^{(m)}$, with the following meaning:

$2^{(s)}$ = 'the insured is disabled with a duration of disability between $s - 1$ and s', for $s = 1, 2, \ldots, m - 1$;

$2^{(m)}$ = 'the insured is disabled with a duration of disability greater than $m - 1'$.

The rationale of this splitting of the disability state is that, as is well known, for different durations since disability inception observed recovery and mortality rates differ; i.e., the splitting allows us to consider select probabilities.

The model requires the following transition probabilities:

$$P_{13}[t], P_{12^{(1)}}[t];$$

$$P_{2^{(s)}1}[t], P_{2^{(s)}3}[t] \quad \text{for } s = 1, 2, \ldots, m;$$

of course the following relations hold:

$$P_{2^{(s)}2^{(s+1)}}[t] = 1 - P_{2^{(s)}1}[t] - P_{2^{(s)}3}[t] \quad \text{for } s = 1, 2, \ldots, m - 1;$$

$$P_{2^{(m)}2^{(m)}}[t] = 1 - P_{2^{(m)}1}[t] - P_{2^{(m)}3}[t].$$

For example, with $m = 3$, the matrix of transition probabilities given by Table 2.1 must be assigned.

As far as recovery is concerned, morbidity experience suggests for any attained age and then for any t:

$$P_{2^{(1)}1}[t] > P_{2^{(2)}1}[t] > \cdots > P_{2^{(m)}1}[t];$$

Table 2.1 Splitting of state 2; one-year transition probabilities

State at time t	State at time $t + 1$				
	1	$2^{(1)}$	$2^{(2)}$	$2^{(3)}$	3
1	$P_{11}[t]$	$P_{12^{(1)}}[t]$	0	0	$P_{13}[t]$
$2^{(1)}$	$P_{2^{(1)}1}[t]$	0	$P_{2^{(1)}2^{(2)}}[t]$	0	$P_{2^{(1)}3}[t]$
$2^{(2)}$	$P_{2^{(2)}1}[t]$	0	0	$P_{2^{(2)}2^{(3)}}[t]$	$P_{2^{(2)}3}[t]$
$2^{(3)}$	$P_{2^{(3)}1}[t]$	0	0	$P_{2^{(3)}2^{(3)}}[t]$	$P_{2^{(3)}3}[t]$
3	0	0	0	0	1

in particular, it is possible to put:

$$P_{2^{(s)}1}[t] > 0 \quad \text{for } s = 1, 2, \ldots, m-1$$
$$P_{2^{(m)}1}[t] = 0;$$

in this case, no recovery is possible after $m - 1$ years.

Remark
This differentiation of disability states, in the framework of a time-discrete Markov model, was originally proposed by Amsler (1968). A pricing procedure based on such a model is used by Dutch insurers; see Gregorius (1993), Wolthuis (1994).

2.4 ACTUARIAL VALUES, PREMIUMS AND RESERVES

In order to illustrate the calculation of actuarial values, premiums and reserves in a time-discrete Markov context, we shall concentrate our attention on a three-state model related to disability annuities. Example 9 in Chapter 1 deals with such a model.

2.4.1 Single premium

Consider an annuity benefit in the case of total disability. Let the policy term be n, let b denote the annual constant amount of disability annuity and let the stopping time (from policy issue) of the annuity payment be n. We assume that the annuity is payable at time h if the insured is then disabled, i.e. is in state 2, for $h = 1, 2, \ldots, n$. Its actuarial value at time 0, with $S(0) = 1$, is given by $ba_{12}(0, n)$, where:

$$a_{12}(0, n) = \sum_{h=1}^{n} v^h P_{12}(0, h). \tag{2.15}$$

In equation (2.15) $v = 1/(1 + i)$ denotes the annual discount factor, and i is the rate of interest.

The **equivalence principle** is satisfied if and only if the single premium is equal to $ba_{12}(0, n)$.

If we assume a time-homogeneous Markov process we simply have

$$P_{12}(0, h) = P_{12}^{(h)}$$

where $P_{12}^{(h)}$ is the $(1, 2)$ entry of the power matrix P^h.

Note that $P_{12}(0, h)$ is the probability that the insured is disabled at time h given that he is active at time 0. This event can be thought as the logical sum of the h events 'the insured is disabled at time h since time $j + 1$, given that he is active at time 0', $j = 0, 1, \ldots, h - 1$. These events are mutually exclusive, so we can write:

$$P_{12}(0,h) = \sum_{j=0}^{h-1} P_{11}(0,j)P_{12}(j,j+1)P_{\underline{22}}(j+1,h). \qquad (2.16)$$

Equation (2.16) can be formally proved via recursion by means of the Chapman–Kolmogorov equations, starting from

$$P_{12}(0,h) = P_{11}(0,h-1)P_{12}(h-1,h) + P_{12}(0,h-1)P_{22}(h-1,h);$$

for the sake of brevity we omit the proof.

Defining

$$\pi(j,h) = P_{11}(0,j)P_{12}(j,j+1)P_{\underline{22}}(j+1,h) \qquad (2.17)$$

we can write

$$a_{12}(0,n) = \sum_{h=1}^{n} v^h \sum_{j=0}^{h-1} \pi(j,h)$$

and then, via simple manipulations

$$a_{12}(0,n) = \sum_{k=1}^{n} \sum_{g=k}^{n} v^g \pi(k-1,g).$$

Finally, from equation (2.17), we obtain the following expression:

$$a_{12}(0,n) = \sum_{k=1}^{n} v^k P_{11}(0,k-1)P_{12}(k-1,k) \sum_{g=k}^{n} v^{g-k} P_{\underline{22}}(k,g) \qquad (2.18)$$

which can be interpreted as follows. The quantity

$$\sum_{g=k}^{n} v^{g-k} P_{\underline{22}}(k,g) = a_{\underline{22}}(k,n)$$

is the actuarial value at time k of a disability annuity of 1 per annum payable up to time n, based on the probability of remaining disabled,

$$P_{\underline{22}}(k,g) = \prod_{h=k}^{g-1} P_{22}(h,h+1); \qquad (2.19)$$

the product $P_{11}(0,k-1)P_{12}(k-1,k)$ represents the probability of being active at time $k-1$, becoming disabled between $k-1$ and k and being disabled at time k. Thus, formula (2.18) is an 'inception-annuity' formula built on the probability of becoming disabled and the probability of remaining disabled, and hence on the actuarial value of a disability annuity. Formula (1.157) in Chapter 1 represents the time-continuous version of this approach for evaluating disability annuities. Conversely, note that formula (2.15) is based on the probability of being disabled; its time-continuous counterpart is given by formula (1.152).

Remark 1

It should be stressed that the structure of the disability benefit we are discussing here is very simple and somewhat unrealistic. In particular, it is assumed that the policy provides payments at integer times h, $h = 1, 2, \ldots, n$ if the insured is then disabled, i.e. at policy anniversaries. It would be more realistic to assume that the annuity is paid (yearly or other than yearly) starting from the time of disability inception (or after a given deferred period). A more realistic set of assumptions will be considered in Chapter 3, when defining a time-continuous context as well as when discussing some approximation formulae. However, note that the assumption of 'payments' at policy anniversaries $(h = 1, 2, \ldots, n - 1)$ if the insured is then surviving and disabled describes the timing of a benefit consisting of waiver during the continuance of disability of the premiums for the life insurance policy which contains the disability clause.

2.4.2 Annual premiums

Given the actuarial value of the disability benefits, i.e. $ba_{12}(0, n)$, calculated at time 0 with $S(0) = 1$, we can determine the annual premiums. Assuming that premiums are paid at integer times k, $k = 0, 1, \ldots, m - 1$ $(m \leq n)$ if the insured is then active, i.e. if $S(k) = 1$, the equivalence principle is satisfied if and only if

$$\sum_{k=0}^{m-1} p_1(k) v^k P_{11}(0, k) = ba_{12}(0, n) \tag{2.20}$$

where $p_1(k)$ denotes the premium at time k.

Note that if we assume a time-homogeneous Markov process we simply have

$$P_{11}(0, k) = P_{11}^{(k)}$$

where $P_{11}^{(k)}$ is the $(1, 1)$ entry of the power matrix P^k (which is the identity matrix if $k = 0$).

Equation (2.20) admits an infinite number of solutions. In particular, if annual premiums are level, we simply have:

$$pa_{11}(0, m) = ba_{12}(0, n) \tag{2.21}$$

where p is the amount of the level annual premium and

$$a_{11}(0, m) = \sum_{k=0}^{m-1} v^k P_{11}(0, k). \tag{2.22}$$

Remark 2

In Chapter 3 we shall illustrate several formulae for evaluating disability benefits, some of which are of the time-discrete type (see in particular sections 3.4.1 and 3.4.2). Moreover, some approximation formulae to determine level annual premiums will be discussed.

2.4.3 Reserves

We shall restrict our attention to prospective reserves, which are defined, for any integer time t and any state $S(t)$, as the actuarial value of future benefits less future premiums. When disability benefits are concerned, we have to define the active reserve and the disabled reserve.

The active reserve is:

$$V_1(t) = ba_{12}(t, n) - pa_{11}(t, m) \quad \text{if } t < m \qquad (2.23a)$$

$$V_1(t) = ba_{12}(t, n) \qquad \qquad \text{if } m \leq t < n \qquad (2.23b)$$

the meaning of $a_{12}(t, n)$ and $a_{11}(t, m)$ coming from a trivial generalization of equations (2.18) and (2.22) respectively.

Conversely, the definition of the disabled reserve is as follows:

$$V_2(t) = ba_{22}(t, n) - pa_{21}(t, m) \quad \text{if } t < m \qquad (2.24a)$$

$$V_2(t) = ba_{22}(t, n) \qquad \qquad \text{if } m \leq t < n \qquad (2.24b)$$

where

$$a_{22}(t, n) = \sum_{k=t+1}^{n} v^k P_{22}(t, k) \qquad (2.25)$$

$$a_{21}(t, m) = \sum_{k=t+1}^{m-1} v^k P_{21}(t, k). \qquad (2.26)$$

It should be stressed that definition (2.25) is consistent with the (usual) convention that the prospective reserves are calculated at a point in time after the payment of the benefit then due but before the receipt of the premium then due.

Note that (2.24a) (as well as (2.24b)) allows for possible recovery and subsequent disability spells; hence, the actuarial value of future premiums is taken into account. In practice, scarcity of data and some computational difficulties often lead to approximation formulae. For example, the following approximation is widely used:

$$V_2(t) = ba_{\underline{22}}(t, n) = b \sum_{k=t+1}^{n} v^k P_{\underline{22}}(t, k). \qquad (2.27)$$

In equation (2.27), only the actuarial value of the current disability annuity is considered.

2.5 EMERGING COSTS; PROFIT TESTING

2.5.1 Some preliminary ideas

According to the traditional approach, actuarial values (i.e. expected present values) of different types of benefits are calculated separately; that is, for each component we consider at the same time the contributions from all the policy duration. In contrast with the so-called **cash flow/ emerging costs** approach, we consider, for each year (or time interval) in turn the contributions from all the different components, all at the same time. This approach allows for a detailed analysis of annual incomes and outgoes.

The cash flow/emerging costs approach leads to **profit testing**, i.e. to the process of adjusting the features of a contract – in particular, the premiums – until the sequence of annual profits satisfies some profit criterion. Thus, profit testing enables the incidence of profit to be taken into account in policy design.

The traditional actuarial methods used in valuing life office products were developed with a view to making calculation procedures as simple as possible. With the more common use of computers, actuaries are able to be more flexible in the nature and types of calculations they perform.

As the present textbook is mostly concerned with calculations of actuarial values in the framework of multiple state models and relevant applications to disability benefits, we do not consider explicitly the algebra of profit emergence; thus, profit testing procedures will not be described. Nevertheless, the concepts of cash flow and emerging costs will be introduced in this section, as they represent the basic ideas leading to profit testing.

2.5.2 Expected cash flow; emerging costs

Consider the annuity benefit described at the beginning of section 2.4. Let the premium function be $p_1(k)$ for $k = 0, 1, \ldots, n-1$, i.e. payable in advance while the insured is in state 1. Then the equivalence principle (see equation (2.20)) is fulfilled if $p_1(k)$ satisfies the following equation:

$$\sum_{k=0}^{n-1} v^k P_{11}(0,k) p_1(k) = b \sum_{k=1}^{n} v^k P_{12}(0,k). \qquad (2.28)$$

We can use the particular example provided by this disability cover to illustrate the adaptation of multiple state models to consideration of cash flows and profit. Some assumptions are needed in order to evaluate cash flows and profit.

First, we assume that the (deterministic) rate of interest earned in the tth year is i_t^*. For notational simplicity and without loss of generality, we can assume $i_t^* = i^*$. Moreover, we have to express a 'realistic' assumption about the transition probabilities, whereas the probabilities used in premium calculation should be 'prudentially' chosen. We will denote with $P_{ij}^*(0, k)$, $P_{ij}^*[k]$ the realistic probabilities. The set of assumptions made in order to evaluate cash flows and profit is often called the **second-order basis** (while the **first-order basis** is the one used in premium calculation).

With the notation of equation (2.6), i.e.

$$P_{ij}^*[z] = P_{ij}^*(z, z + 1),$$

we define the expected cash flow for year t of the policy per policy 'alive' (i.e. in state 1 or 2) at the start of year t as follows:

$$CF_1(t) = p_1(t-1)(1+i^*) - bP_{12}^*[t-1] \quad \text{if the policy holder is in state 1 at the start of the year} \quad (2.29a)$$

$$CF_2(t) = \qquad\qquad\qquad - bP_{22}^*[t-1] \quad \text{if the policy holder is in state 2 at the start of the year} \quad (2.29b)$$

for $t = 1, 2, \ldots, n$. This is the amount of money which is expected to emerge at the end of year t for each policy 'alive' at the start of the year.

We then define the expected cash flow for year t per policy issued at time 0, or **emerging costs** for year t, as follows

$$EC(t) = P_{11}^*(0, t-1)CF_1(t) + P_{12}^*(0, t-1)CF_2(t)$$

$$= P_{11}^*(0, t-1)[p_1(t-1)(1+i^*) - bP_{12}^*[t-1]]$$

$$+ P_{12}^*(0, t-1)[-bP_{22}^*[t-1]] \qquad (2.30)$$

for $t = 1, 2, \ldots, n$, using the fact that the policyholder is in state 1 at the issue of the policy.

Assume $i^* = i$ and $P_{ij}^*(0, k) = P_{ij}(0, k)$ for any i, j, k. It is then straightforward to show that the equation determining $p_1(k)$, obtained by the equivalence principle, which is of the form of equation (2.28), is identical to

$$\sum_{t=1}^{n}(1+i)^{-t}EC(t) = 0. \qquad (2.31)$$

Thus, the conventional premium equation derived from the equivalence principle may be written alternatively as the 'present value of the emerging costs equals zero' (the present value of the emerging costs being calculated with the first order basis).

It is straightforward to include expenses and other types of benefit in the definitions of $CF_i(t)$ and $EC(t)$.

2.5.3 Expected profit

The set of $CF_i(t)$ for a policy does not represent expected profit to the insurer since funds must be set aside to meet the expected outgo in future years (i.e. in year u, $u > t$), for example for paying the benefits. This would be indicated by $CF_i(t)$ being negative for certain t and/or for certain states. But, once suitable reserves have been specified and set up, the balance of the cash flow in any year can be regarded as profit to the insurer. So, in any year the profit depends both on the cash flow and the reserves calculated (and the underlying assumptions).

We define $PR_j(t)$ to be the profit expected to emerge at the end of year t per policy in state j at time $t - 1$. Then, for $j = 1, 2$,

$$PR_j(t) = CF_j(t) + (1 + i^*)V_j(t - 1) - \sum_{k=1}^{2} P_{jk}^*[t - 1]V_k(t). \qquad (2.32)$$

It is usual to adopt the convention that the prospective reserves are calculated at point of time after the payment of the benefit then due but before the receipt of the premium then due, so that

$$V_j(0) = V_j(n) = 0.$$

In general, the summation on the right-hand side of equation (2.32) is over permissible values of k.

Different types of policy have different characteristic forms of $PR_j(t)$. Thus, the incidence of expenses often means that $PR_j(1) < 0$. Then the insurer must allocate capital to cover this initial 'loss', often called the 'new business strain'. This capital might be provided by the shareholders or by other groups of policyholders. The insurer will need to earn an adequate rate of return on this capital so invested and this return will derive from later non-negative values of the profit emerging, $PR_j(t)$.

Analysis of the vectors $PR_j(t)$ is most commonly referred to as **profit testing**. It provides a flexible and powerful tool for considering the profitability of policies with complex designs. We note that the components of the profit vectors depend on the parameter values and assumptions underlying the calculations and so do not necessarily represent the actual profit which the insurer will receive, but rather the profits expected if the actual experience turns out to match exactly the assumptions made.

2.6 REFERENCES AND SUGGESTIONS FOR FURTHER READING

In this section also we only quote papers dealing with general aspects of multiple state modelling in life insurance and related fields. Studies particularly devoted to applications of multiple state modelling to disability annuities, long-term care insurance, critical illness cover, etc., will be quoted in the relevant sections of the following chapters.

While there are a number of theoretical papers dealing with actuarial multiple state models in a time-continuous context, the literature on time-discrete models is very scanty. Probably, this bears witness to the validity of the time-continuous approach, at least as far as the actuarial field is concerned. Conversely, we will see in Chapter 3 that a very extensive application oriented literature is devoted to time-discrete pricing and reserving methods.

The first paper dealing with the insurances of the person within a time-discrete Markov context is probably due to Franckx (1963). In this paper life and non-life actuarial problems are treated within a Markov framework. Particular attention is devoted to life and disability insurance.

Actuarial problems concerning annuities, life assurance and disability covers are discussed within a time-discrete Markov context by Daboni (1964). Amsler (1968, 1988) describes the Markov structure of life and disability insurance both in time-discrete and time-continuous terms, also discussing algorithmic aspects. Substates of the disability state are also considered, in order to express the durational effect on the probabilities of transitions from the disability state.

A more recent contribution in the field of Markov models for life and other contingencies is due to Gatenby and Ward (1994).

For further discussion of the cash flow methodology for simple and multiple decrement models and multiple state models, readers are referred to Chadburn *et al.* (1995), and for further discussion of profit testing, albeit in the context of life insurance, readers can consult Hare and McCutcheon (1991).

3

Disability insurance

3.1 TYPES OF BENEFITS

A short description of some types of disability policies follows. For a more detailed illustration, the reader should consult the references quoted at the end of each subsection.

3.1.1 United Kingdom: permanent health insurance (PHI)

Individual PHI policies are designed to provide a weekly or monthly income to an individual if he/she is prevented by sickness or injury from working. In this sense, PHI policies provide disability income protection.

Because the occurrence of a disability claim may depend on the characteristics of the insured other than those related to the disability itself, there is the potential for moral hazard. The main reason for moral hazard arising is that, in the event of disablement, the insured's total income may increase or decrease to a significant effect.

Hence, the size of the income benefit needs to be considered by the underwriter at the time of application in relation to the applicant's current financial situation and the level of benefits the applicant may expect to receive in the event of disablement from social security, company pension plans or other insurances. The policy conditions usually include a clause to limit the income provided by the policy if this is excessive (according to a pre-specified formula) in relation to the insured's normal level of income (allowing for tax, other sources of income, etc.).

Policies normally include a deferred period so that the benefit will not begin to be payable until the sickness has lasted a certain minimum period (the deferred period; usually four, 13 or 26 weeks). As PHI policies are usually effected to supplement the sickness benefit available from an employer or the benefit payable from the state (e.g. in the UK, the National Insurance scheme), the deferred period chosen by the policyholder will tend to reflect the length of time after which these benefits reduce or (in the case of benefits from an employer) cease. Obviously, the longer the deferred period, the cheaper the cover, and hence the lower the premium.

Policies are usually for a fixed term, usually ceasing at age 65 for males or age 60 for females.

Once the insurance company has offered formally to provide the necessary cover and the first premium has been paid, the company cannot cancel the policy, so long as the policyholder obeys the policy conditions (hence the name 'permanent' health insurance).

Under the most common type of PHI policy, a weekly or monthly income commences to be paid to the policyholder when he/she has been sick for longer than the deferred period. The benefit continues to be paid until the policyholder recovers or dies or until the age at which the policy term ceases. Because the insurer cannot cancel the policy, a policyholder who is sick permanently, or indefinitely, will receive the benefit throughout, until one of the above events occurs. With most policies, the premiums are waived while the income benefit is being paid.

Some policies pay a benefit of a fixed level amount while other provide a benefit which increases in some way in an effort to protect the policyholder from the effects of inflation. There are various methods by which increases in benefits are provided, some of which are more effective than others. Policies are usually designed so that increasing benefits are matched by increasing premiums, whether annual premiums or recurrent single premiums.

Another feature in policy design is a benefit level that reduces with duration of the sickness claim. This is designed to encourage a return to work. The availability of the PHI benefit may lengthen the duration of sickness because of its effect on the minimum acceptable salary that would 'entice' the sick individual to return to gainful employment (i.e. the so-called reservation wage).

Policies in the UK follow a number of distinct designs:

1. conventional policies where the terms are guaranteed throughout the life of the policy term;
2. short-term renewable policies;
3. unit-linked policies;
4. 'keyman' policies which cover the risks of a key employee disabled and where the income benefit is payable only for one or two years.

Such policies are offered on an individual basis. Group PHI policies (for a group of employees) generally involve recurrent single premiums (rather than regular annual premiums) and level or escalating benefits are also provided.

For further information, readers should consult Clark and Baker (1992), Mackay (1993) or Sanders and Silby (1988).

3.1.2 USA

In the USA, the disability income policy is often called 'loss of time' or 'loss of income' or 'long-term disability'. It provides payments when an insured is unable to work because of injury or sickness. Policy designs are similar to those in the UK, although a different nomenclature is used; for example, '(benefit) elimination' or 'waiting' period are commonly used, instead of 'deferred period'. It is also common to include a maximum benefit period in the event of the total disability of the insured policyholder. Benefits are normally paid monthly as fixed payments while the insured is disabled. As in the UK, this insurance is available on both an individual and a group basis.

In group coverages the benefit amount is related directly to the insured's earnings, at least in part. In individual coverages the relationship is much more loose and even then only exists at time of policy issue. Benefits may be co-ordinated with social security, decreasing if social security benefits are received.

Policies are offered on a stand-alone basis and as riders to other types of long term insurance policy; however, the second possibility is rare nowadays.

For further information readers should consult Black and Skipper (1987), O'Grady (1988), and Minor (1968). Partial disability benefits related to the so-called 'residual disability' are dealt with by Miller (1984).

3.1.3 The Netherlands

In the Netherlands two main types of disability cover are sold: the 'first year risk cover' (A-cover) and the 'after-first year risk cover' (B-cover).

The A-cover policy conditions include a deferment period of between seven days and six months, and a maximum period of annuity payment equal to one year. The insured is entitled to benefits when he is at least 25% disabled and unable to perform the duties connected with his occupation.

The B-cover policy conditions include a one-year deferment period. The disability annuity stops in case of death or recovery or when the policy term ceases. The insured is entitled to benefits when he is at least 25% disabled and unable to perform the duties which meet his abilities and capacities and which can be reasonably demanded from him in view of his education and former activities. The insured amounts can be indexed according to a fixed rate: 3, 4 or 5%.

For further information, readers are referred to Gregorius (1993).

3.1.4 Germany, Austria, Switzerland

In Germany, both disability riders and stand-alone disability covers are offered. The disability definition is as follows.

> The insured shall be deemed to be suffering from total disability if, as a result of illness or bodily injury which must be medically demonstrable, he is expected to be permanently unable to practise his occupation or carry out another activity which can be performed by reason of his training and experience and which corresponds to his social standing. The insured shall be deemed to be suffering from partial disability if the prerequisites specified above are expected to be met permanently to a certain degree only. If, as a result of illness or bodily injury which must be medically demonstrable, the insured has for an uninterrupted period of six months been totally or partially unable to practise his occupation or carry out another activity which can be performed by reason of his training and experience and which corresponds to his social standing, the continuation of his condition shall be deemed to be total or partial disability.

Generally, insurers pay the whole benefit insured in the case of a disability greater than or equal to 50%.

In Austria, disability covers are only offered in the form of riders to life policies. The definition of disability is similar to that used in Germany.

In Switzerland disability insurance is sold both in the form of a rider to a life policy and as stand-alone cover. Both total disability and partial disability are covered, the latter with correspondingly reduced benefits. The benefits are scaled according to the degree of disablement. Full benefits are paid if the degree is greater than or equal to 66.6%, whereas none are paid if the degree is less than 25%. The definition of disability is as follows.

> Disability is the total or partial inability of the insured due to illness or accident to engage in his own occupation or in any other occupation that he may reasonably be expected to undertake in view of his situation in life, his knowledge and his abilities.

For further information on disability insurance in Germany, Austria and Switzerland, readers should consult Segerer (1993).

Remark

The emphasis of this book is on the actuarial models needed for the proper scientific pricing of and reserving for disability insurance. We pay less attention to the essential practical aspects of writing disability insurance business such as underwriting, claims management and control, marketing and experience monitoring. We fully accept that these elements play an important role in the emerging experience and that they cannot be ignored when considering pricing or reserving in a practical context where it would be imperative to ensure that any premiums charged and reserves established are consistent with these other elements. Space does not permit us to pursue these other dimensions but readers are referred to one of the texts cited earlier, to Miller (1976), and to Goford (1985) for a presentation (albeit in a life insurance framework) of these components within the context of a 'control cycle'.

3.2 GENERAL ACTUARIAL ASPECTS CONCERNING A DISABILITY ANNUITY

3.2.1 Notation

In this and the following chapters a notation is used which differs from the one adopted in the previous chapters. Notation in Chapter 1 and 2 comes from probability theory and seems to be the new actuarial standard, at least as far as 'theoretical' actuarial mathematics is concerned.

However, some problems arise when turning to practical implementations, and in particular when considering implementations based on time-discrete models. Indeed, in disability annuities actuarial practice the so-called 'Hamza notation' is very often adopted. Moreover, it is rather easy to extend this notation in order to write formulae relating to benefits other than disability annuities, such as critical illness cover, or long-term care insurance.

So, we will use a Hamza-like notation in which the letters a, i, d denote the states 'active', 'invalid', 'dead'. Some extensions (for example: $i_1, i_2, \ldots$) allow for specifying the severity of disability, the durational effect, etc. Age will be explicitly indicated; thus, x indicates the entry age in an aggregate model, while $[x]$ denotes the entry age when issue-selection is taken into account.

Hence, for an insured aged x at policy issue, assuming that state 1 means 'active' and state 2 'invalid', some correspondences are illustrated by the following examples:

$$P_{11}(0,t) \quad \rightarrow \quad {}_tp_x^{aa}$$

$$\rightarrow \quad {}_tp_{[x]}^{aa} \quad \text{(in case of issue-selection)}$$

$$P_{11}(z,t) \quad \rightarrow \quad {}_{t-z}p_{x+z}^{aa}$$

$$P_{\underline{11}}(0,t) \quad \rightarrow \quad {}_tp_x^{\overline{aa}}$$

$$P_{12}(0,t) \quad \rightarrow \quad {}_tp_x^{ai}$$

$$\mu_{12}(t) \quad \rightarrow \quad \mu_{x+t}^{ai}$$

$$\bar{a}_{11}(0,n) \quad \rightarrow \quad \bar{a}_{x:n\rceil}^{aa}$$

$$\bar{a}_{11}(0,\infty) \quad \rightarrow \quad \bar{a}_x^{aa}$$

$$\bar{a}_{\underline{22}}(t,n) \quad \rightarrow \quad \bar{a}_{x+t:n-t\rceil}^{\underline{ii}}.$$

As far as an inception-select model is adopted, we write for instance (see equation (1.64)):

$$P_{\underline{22}}(z, t, r) \quad \longrightarrow \quad {}_{t-z}\acute{p}^{\overset{ii}{}}_{x+z,r}$$

$$\mu_{21}(t, r) \quad \longrightarrow \quad \mu^{ia}_{x+t,r}$$

and $\bar{a}^{\overset{ii}{}}_{[y-u]+u:x+n-y]} (x \leq y - u; y \leq x + n)$ to denote the actuarial value of a disability annuity to be paid as long as the insured is in the disability state, until age $x + n$. Other symbols shall be defined when concerned.

3.2.2 The time-continuous model

Let us describe the development of an individual insurance policy which pays out a continuous annuity of dt units in the interval $[t, t + dt)$, if the insured is disabled. Let x be the age of the insured at policy issue. Let $S(x + t)$ represent the random state occupied by the policyholder at age $x + t$; for any t, $t > 0$, the possible realizations of $S(x + t)$ are: a ('active' or 'healthy'), i ('invalid' or 'ill' or 'disabled'), d ('dead'). For example, the statement '$S(x + t) = a$' means 'the policyholder is in the state a at age $x + t'$. We assume $S(x) = a$. The graph of Fig. 3.1 illustrates the set of states and the set of possible transitions between states.

Assuming a lifetime duration for the policy cover, the random present value Y_x at time 0 of the benefits paid by the insurer is given by:

$$Y_x = \int_0^{+\infty} v^u I_{\{S(x+u)=i\}} \, du \tag{3.1}$$

where $I_{\{S(x+u)=i\}}$ denotes the indicator of the event $S(x + u) = i$, and v is the annual discount factor. Let

$$\phi(x, u) = \Pr\{S(x + u) = i \mid S(x) = a\}. \tag{3.2}$$

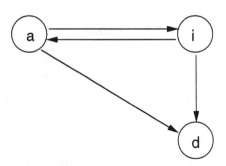

Figure 3.1 Set of states and set of transitions

Then, the actuarial value, i.e. the net single premium $\bar{a}_x^{ai}$ (denoted according to the usual actuarial notation), is given by

$$\bar{a}_x^{ai} = E(Y_x \mid S(x) = a) = \int_0^{+\infty} v^u \phi(x, u) \, du. \tag{3.3}$$

As discussed in section 3.1, it is customary for policy conditions to permit the benefits to be paid out if and only if certain conditions concerning the duration of the disability are fulfilled. Let us define:

$\phi^\Gamma(x, u) =$ probability that the insured (healthy at age x)
is disabled and the annuity is payable at age $x + u$,
according to policy conditions Γ; (3.4)

of course, we have

$$\phi^\Gamma(x, u) \le \phi(x, u). \tag{3.5}$$

The set of policy conditions can be formally represented by a set of five parameters

$$\Gamma = [n_1, n_2, f, m, r]$$

where:

- (n_1, n_2) denotes the insured period (an annuity is payable if disability inception time belongs to this interval); for example:
  $n_1 = c =$ waiting period (from policy issue);
  $n_2 = n =$ policy term;
- f is the deferred period (from disability inception);
- m is the maximum number of years of annuity payment (from disability inception);
- r is the stopping time (from policy issue) of annuity payment; for example, if $\xi =$ retirement age, then we may have $r = \xi - x$.

Note that if $m = \infty$ the longest period of disability annuity payment is implicitly given by r.

For instance, we have:

$$\phi^{[0,\infty,0,\infty,\infty]}(x, u) = \phi(x, u)$$

$$\phi^{[0,n,0,\infty,n]}(x, u) = \begin{cases} \phi(x, u) & \text{if } u < n \\ 0 & \text{if } u \ge n. \end{cases}$$

Thus, the net single premium for $\Gamma = [0, n, 0, \infty, n]$ is given by

$$\bar{a}_{x,\Gamma}^{ai} = \int_0^{+\infty} v^u \phi^{[0,n,0,\infty,n]}(x, u) \, du = \int_0^n v^u \phi(x, u) \, du \tag{3.6}$$

and, in this case, it is denoted by $\bar{a}_{x:\overline{n}|}^{ai}$ according to the usual actuarial notation.

3.2.3 The Markov assumption

First, we assume that the stochastic process $\{S(x + t); t \geq 0\}$ is a time-continuous, time-inhomogeneous, three-state Markov chain (see definitions (1.5) and (1.8)). Let $_tp_y^{gh}$ denote the transition probabilities:

$$_tp_y^{gh} = \Pr\{S(y + t) = h \mid S(y) = g\}; \quad h = a, i, d; g = a, i. \tag{3.7}$$

The transition intensities are defined as

$$\mu_y^{gh} = \lim_{t \to 0} \frac{_tp_y^{gh}}{t}; \quad h = a, i, d; g = a, i; h \neq g. \tag{3.8}$$

These limits are assumed to exist for all relevant y and the intensities are assumed to be integrable on compact intervals.

Moreover, we denote the occupancy probabilities as follows:

$$_tp_y^{\overline{hh}} = \Pr\{S(y + u) = h \text{ for all } u \in [0, t] \mid S(y) = h\}; \quad h = a, i. \tag{3.9}$$

The transition probabilities satisfy the Chapman–Kolmogorov relation (see equation (1.12)) and the Kolmogorov forward and backward differential equations (see equations (1.19) and (1.20)). Refer to Example 9 in Chapter 1 as far as finding transition probabilities from intensities is concerned).

In actuarial calculations concerning a PHI policy, the following probability is often required:

$$_tp_y^{ai}(\tau) = \Pr\{S(y + u) = i \text{ for all } u \in [t - \tau, t] \mid S(y) = a\} \tag{3.10}$$

where $0 \leq \tau \leq t$. Of course we have

$$_tp_y^{ai}(0) = {}_tp_y^{ai} \tag{3.10'}$$

and

$$_tp_y^{ai}(\tau) = 0 \quad \text{when } \tau \geq t. \tag{3.10''}$$

For $0 \leq \tau \leq t$, the following expression can be derived:

$$_tp_y^{ai}(\tau) = \int_0^{t-\tau} {}_up_y^{\overline{aa}}\, \mu_{y+u}^{ai}\, {}_{t-u}p_{y+u}^{\overline{ii}}\ du \tag{3.11}$$

(which can be easily interpreted by direct reasoning).

3.2.4 The semi-Markov assumption

As pointed out in section 1.6 (see Example 9), more realistic models can be built considering the dependence of the intensities of recovery and mortality of disabled lives on the time spent in the disability state since latest transition to that state. This naturally leads to a semi-Markov process, in which the above-mentioned intensities are functions of the attained age and the duration of disability as well. In sections 3.3.2 and 3.3.3 we will illustrate some practical implementations in which intensities of this type are involved.

The duration of presence in the disability state can also be handled in a different way, i.e. by splitting the disability state into a number of states and hence using a number of intensities (each intensity being a function of the attained age only), reflecting the durational effect (see section 1.7). In section 3.4.2 we will discuss a practical implementation in which the durational effect is handled by splitting the disability state.

For the sake of simplicity, in this section we shall restrict our attention to Markov models based on the space state $\mathscr{S} = \{a, i, d\}$.

3.2.5 Modelling policy conditions

Let us return to the probabilities $\phi(x, t)$ and, more generally, $\phi^{\Gamma}(x, t)$ defined in section 3.2.2. The Markovian model allows for a straightforward expression of these probabilities. As already pointed out, no durational effect is considered at this time.

In particular, we obviously have:

$$_tp_x^{ai} = \phi(x, t) = \phi^{[0,\infty,0,\infty,\infty]}(x, t). \tag{3.12}$$

The following examples concern more realistic policy conditions.

$$\phi^{[c,\infty,0,\infty,\infty]}(x, t) = \begin{cases} 0 & \text{if } t \le c \\ _tp_x^{ai} - {_tp_x^{ai}}(t - c) & \text{if } t > c \end{cases} \tag{3.13}$$

$$\phi^{[0,\infty,f,\infty,\infty]}(x, t) = \begin{cases} 0 & \text{if } t \le f \\ _tp_x^{ai}(f) & \text{if } t > f \end{cases} \tag{3.14}$$

$$\phi^{[0,n,0,\infty,\infty]}(x, t) = \begin{cases} _tp_x^{ai} & \text{if } t < n \\ _tp_x^{ai}(t - n) & \text{if } t \ge n \end{cases} \tag{3.15}$$

$$\phi^{[0,\infty,0,m,\infty]}(x, t) = {_tp_x^{ai}} - {_tp_x^{ai}}(m) \tag{3.16}$$

$$\phi^{[0,n,0,m,\infty]}(x, t) = \begin{cases} _tp_x^{ai}(\max(0, t - n)) - {_tp_x^{ai}}(m) & \text{if } t < n + m \\ 0 & \text{if } t \ge n + m \end{cases} \tag{3.17}$$

$$\phi^{[0,n,0,\infty,n]}(x, t) = \begin{cases} _tp_x^{ai} & \text{if } t < n \\ 0 & \text{if } t \ge n \end{cases} \tag{3.18}$$

$$\phi^{[0,n,0,m,n]}(x, t) = \begin{cases} _tp_x^{ai} - {_tp_x^{ai}}(m) & \text{if } t < n \\ 0 & \text{if } t \ge n \end{cases} \tag{3.19}$$

$$\phi^{[0,n,0,\infty,r]}(x,t) = \begin{cases} {}_tp_x^{ai} & \text{if } t < n \\ {}_tp_x^{ai}(t-n) & \text{if } n \le t < r \\ 0 & \text{if } t \ge r. \end{cases} \qquad (3.20)$$

Now, let us go back to the probability ${}_tp_y^{ai}$, which can be expressed in the following form (see equation (1.40) in section 1.5):

$$_tp_y^{ai} = \int_0^t {}_up_y^{aa}\,\mu_{y+u}^{ai}\cdot {}_{t-u}p_{y+u}^{ii}\;du. \qquad (3.21)$$

From equation (3.3) we get:

$$\bar{a}_x^{ai} = \int_0^{+\infty} \phi(x,t)v^t\;dt = \int_0^{+\infty} {}_tp_x^{ai}v^t\;dt \qquad (3.22)$$

and equation (3.21) yields:

$$\bar{a}_x^{ai} = \int_0^{+\infty}\int_0^t v^t\, {}_up_x^{aa}\,\mu_{x+u}^{ai}\cdot {}_{t-u}p_{x+u}^{ii}\;du\;dt. \qquad (3.23)$$

Changing the order of integration, we finally obtain:

$$\bar{a}_x^{ai} = \int_0^{+\infty} {}_up_x^{aa}\,\mu_{x+u}^{ai}v^u\left[\int_u^{+\infty} {}_{t-u}p_{x+u}^{ii}v^{t-u}\;dt\right]du. \qquad (3.24)$$

Note that the integral in brackets is the present expected value at age $x+u$ of a continuous annuity payable to a disabled insured until recovery or death. The following notation is usually adopted:

$$\bar{a}_y^{ii} = \int_0^{+\infty} {}_zp_y^{ii}v^z\;dz. \qquad (3.25)$$

It must be stressed that formula (3.22) is based on the probability of **being** disabled (${}_tp_x^{ai}$), whilst (3.24) is based on the probabilities of **becoming** ($\mu_{x+u}^{ai}\;du$) and **remaining** (${}_{t-u}p_{x+u}^{ii}$) disabled. In sections 3.3 and 3.4 we shall illustrate how the most common calculation procedures used in the actuarial practice can be easily interpreted and discussed bearing in mind the two approaches.

A simple graphical representation provides a powerful tool for clarifying some aspects of the calculation of single premiums corresponding to various policy conditions Γ. Figure 3.2 shows a possible development of disability for an insured life. The insured moves on the horizontal axis as long as he remains healthy; at point R in the diagram he becomes disabled and hence moves upward until recovery (segment RS); then he is healthy again (VW), becomes disabled and dies (WZ).

The set of all possible 'health stories' concerning an individual is represented by the shaded region in Fig. 3.3.

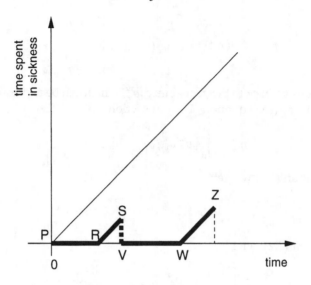

Figure 3.2 A 'health story'.

The single premium $\bar{a}_x^{ai}$ can be interpreted as a measure associated with the shaded region of Fig. 3.3. To prove this, let us return to formula (3.23) and consider the variable $z = t - u$, i.e. the time spent in sickness. This variable is represented on the vertical axis of Figs 3.2 and 3.3. With a change of variable in the inner integral, that is replacing u by z, formula

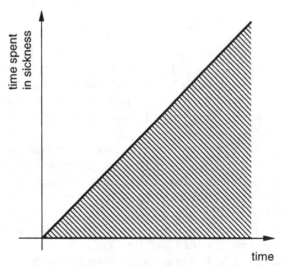

Figure 3.3 Set of all possible 'health stories'.

(3.23) becomes

$$\bar{a}_x^{ai} = \int_0^{+\infty} \int_0^t v^t \,_{t-z}p_x^{aa} \mu_{x+t-z}^{ai} \cdot \,_z p_{x+t-z}^{ii} \, dz \, dt \qquad (3.26)$$

from which we see that the double integral is evaluated over the shaded region.

This result can be extended to more realistic policy conditions. First, we can observe that for any given combination Γ of policy conditions, the relevant set of possible 'health stories' is represented by a particular region which is a subset of the region shaded in Fig. 3.3. Figures 3.4a to 3.4h provide some examples.

Then, any single premium can be interpreted as the value of an integral over the relevant particular region. To prove this, let us consider the function ψ defined as follows:

$$\psi(x, u, t) = \,_u p_x^{aa} \mu_{x+u}^{ai} \cdot \,_{t-u} p_{x+u}^{ii} \qquad (3.27)$$

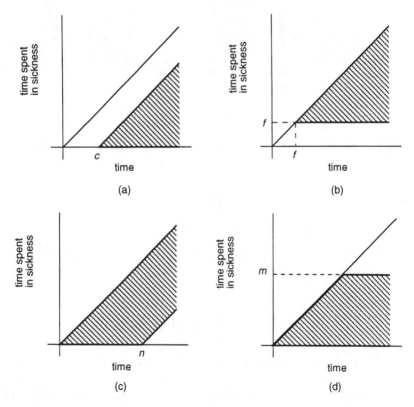

Figure 3.4 (a) $\Gamma = [c, \infty, 0, \infty, \infty]$. (b) $\Gamma = [0, \infty, f, \infty, \infty]$. (c) $\Gamma = [0, n, 0, \infty, \infty]$. (d) $\Gamma = [0, \infty, 0, m, \infty]$.

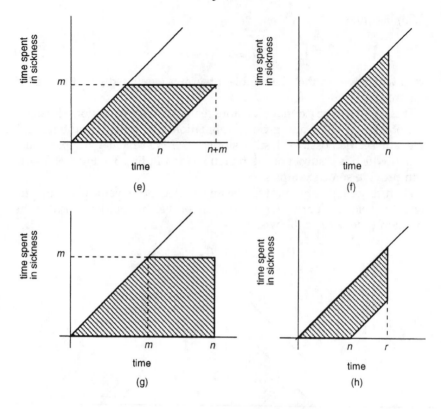

Figure 3.4 (e) $\Gamma = [0, n, 0, m, \infty]$. (f) $\Gamma = [0, n, 0, \infty, n]$. (g) $\Gamma = [0, n, 0, m, n]$. (h) $\Gamma = [0, n, 0, \infty, r]$.

or

$$\psi(x, t - z, t) = {}_{t-z}p_x^{aa}\,\mu_{x+t-z}^{ai} \cdot {}_{z}p^{\overset{ii}{x+t-z}}. \tag{3.28}$$

Hence, for $\Gamma = [0, n, 0, \infty, \infty]$, we get:

$$
\begin{aligned}
\bar{a}_{x,\Gamma}^{ai} &= \int_0^n {}_tp_x^{ai}v^t\,dt + \int_n^{+\infty} {}_tp_x^{ai}(t - n)v^t\,dt \\
&= \int_0^n \int_0^t \psi(x, u, t)v^t\,du\,dt + \int_n^{+\infty}\int_0^n \psi(x, u, t)v^t\,du\,dt \\
&= \int_0^n \int_0^t \psi(x, t - z, t)v^t\,dz\,dt + \int_n^{+\infty}\int_{t-n}^t \psi(x, t - z, t)v^t\,dz\,dt \quad (3.29)
\end{aligned}
$$

i.e. the integration is over the region shaded in Fig. 3.4c. Formula (3.11) has been used to express the probability ${}_tp_x^{ai}(t - n)$ in terms of a double integral of the function $\psi(x, u, t)$.

For $\Gamma = [0, \infty, 0, m, \infty]$, we obtain:

$$\bar{a}^{ai}_{x,\Gamma} = \int_0^{+\infty} {}_t p_x^{ai} v^t \, dt - \int_m^{+\infty} {}_t p_x^{ai}(m) v^t \, dt$$

$$= \int_0^{+\infty} \int_0^t \psi(x, u, t) v^t \, du \, dt - \int_m^{+\infty} \int_0^{t-m} \psi(x, u, t) v^t \, du \, dt$$

$$= \int_0^{+\infty} \int_0^t \psi(x, t - z, t) v^t \, dz \, dt - \int_m^{+\infty} \int_m^t \psi(x, t - z, t) v^t \, dz \, dt \quad (3.30)$$

i.e. the integration is performed over the region shaded in Fig. 3.4d.

A case of special interest (as we will see later) is given by $\Gamma = [0, n, 0, \infty, r]$, with $r > n$. We have:

$$\bar{a}^{ai}_{x,\Gamma} = \int_0^n {}_t p_x^{ai} v^t \, dt + \int_n^r {}_t p_x^{ai}(t - n) v^t \, dt$$

$$= \int_0^n \int_0^t \psi(x, u, t) v^t \, du \, dt + \int_n^r \int_0^n \psi(x, u, t) v^t \, du \, dt$$

$$= \int_0^n \int_0^t \psi(x, t - z, t) v^t \, dz \, dt + \int_n^r \int_{t-n}^t \psi(x, t - z, t) v^t \, dz \, dt \quad (3.31)$$

i.e. the integration is performed over the region shaded in Fig. 3.4h. The single premium for the case $r = n$ (see Fig. 3.4f) can be immediately obtained from equation (3.31).

Similar results can be obtained for the other cases.

3.2.6 Periodic premiums; reserves

Let us refer to an insurance policy providing a continuous disability annuity of 1 per annum according to a given set of policy conditions Γ. Let $\bar{a}^{ai}_{x,\Gamma}$ denote the actuarial value of benefits. As is well known, $\bar{a}^{ai}_{x,\Gamma}$ also represents the single premium fulfilling the equivalence principle. When periodic premiums are involved, it is quite natural to assume that premiums are paid on a continuous basis as we are reasoning within a time-continuous context.

Let us suppose that premiums are paid when the contract stays in state a (i.e. when the insured is active) and not in state i (when disabled), as premiums are usually waived during disability spells. Let $\bar{P}_x(u)$ denote the instantaneous premium rate. Then, the equivalence principle is fulfilled if the function $\bar{P}_x(u)$ satisfies the following equation:

$$\int_0^{+\infty} v^u \, {}_u p_x^{aa} \bar{P}_x(u) \, du = \bar{a}^{ai}_{x,\Gamma}. \quad (3.32)$$

Particular premium functions can be obtained when policy conditions are given. For example, let $\Gamma = [0, n, 0, \infty, n]$. In this case we have (from equation (3.31) with $r = n$ and from equation (3.27)):

$$\bar{a}^{ai}_{x,\Gamma} = \int_0^n \int_0^t {}_u p^{aa}_x \mu^{ai}_{x+u} \cdot {}_{t-u}p^{ii}_{x+u} v^t \, du \, dt. \tag{3.33}$$

After some manipulations we obtain:

$$\bar{a}^{ai}_{x,\Gamma} = \int_0^n {}_u p^{aa}_x \mu^{ai}_{x+u} v^u \left[\int_u^n {}_{t-u}p^{ii}_{x+u} v^{t-u} \, dt \right] du$$

$$= \int_0^n {}_u p^{aa}_x \mu^{ai}_{x+u} v^u \bar{a}^{ii}_{x+u:n-u\rceil} \, du. \tag{3.34}$$

Assuming a constant premium rate, $\bar{P}_x$, over $[0, m)$ with $m \leq n$, we have:

$$\bar{P}_x = \frac{\int_0^n {}_u p^{aa}_x \mu^{ai}_{x+u} v^u \bar{a}^{ii}_{x+u:n-u\rceil} \, du}{\int_0^m v^u \, {}_u p^{aa}_x \, du} = \frac{\bar{a}^{ai}_{x,\Gamma}}{\bar{a}^{aa}_{x:m\rceil}}. \tag{3.35}$$

Now, let us consider $\Gamma = [0, n, 0, \infty, r]$, with $r > n$. In this case we have (from equations (3.31) and (3.27)):

$$\bar{a}^{ai}_{x,\Gamma} = \int_0^n \int_0^t {}_u p^{aa}_x \mu^{ai}_{x+u} \cdot {}_{t-u}p^{ii}_{x+u} v^t \, du \, dt$$

$$+ \int_n^r \int_0^n {}_u p^{aa}_x \mu^{ai}_{x+u} \cdot {}_{t-u}p^{ii}_{x+u} v^t \, du \, dt. \tag{3.36}$$

Finally, after some manipulations we find:

$$\bar{a}^{ai}_{x,\Gamma} = \int_0^n {}_u p^{aa}_x \mu^{ai}_{x+u} v^u \left[\int_u^r {}_{t-u}p^{ii}_{x+u} v^{t-u} \, dt \right] du$$

$$= \int_0^n {}_u p^{aa}_x \mu^{ai}_{x+u} v^u \bar{a}^{ii}_{x+u:r-u\rceil} \, du. \tag{3.37}$$

The expression for the constant premium rate, $\bar{P}_x$, over $[0, m)$ with $m \leq n$, is straightforward.

Disability insurance can also be financed by discrete-time premium mechanisms. In this case, for a given set of times $t_1, t_2, \ldots, t_m$, a sequence $P_x(t_1), P_x(t_2), \ldots, P_x(t_m)$ must be assessed in order to fulfil the equivalence principle. Hence:

$$\sum_{h=1}^m v^{t_h} {}_{t_h}p^{aa}_x P_x(t_h) = \bar{a}^{ai}_{x,\Gamma}. \tag{3.38}$$

In actuarial practice discrete-time premiums are usually calculated as natural premiums. Section 3.3.5 is specifically devoted to this approach, which is of particular interest when group disability insurance is concerned.

As is well known, the prospective reserve at time t is defined as the actuarial value of future benefits less the actuarial value of future premiums, given the state of the policy at time t. Thus, we have to define an **active reserve** as well as a **disabled reserve**. Prospective reserves can be formally expressed when policy conditions are stated.

For example, assume $\Gamma = [0, n, 0, \infty, n]$. In the case of a single premium, the active reserve is given by:

$$\bar{V}^a_{x+t,\Gamma} = \bar{a}^{ai}_{x+t,\Gamma} = \int_0^{n-t} {}_u p^{ai}_{x+t} v^u \, du$$

$$= \int_0^{n-t} {}_u p^{aa}_{x+t} \mu^{ai}_{x+t+u} v^u \bar{a}^{ii}_{x+t+u:\overline{n-t-u}|} \, du. \tag{3.39}$$

The disabled reserve (assuming a non-select stochastic model) is given by:

$$\bar{V}^i_{x+t,\Gamma} = \int_0^{n-t} {}_u p^{ii}_{x+t} v^u \, du. \tag{3.40}$$

Let us now assume that a continuous premium is paid over $[0, m)$ at an instantaneous constant rate $\bar{P}_x$ if the insured is active. In this case, if $t < m$ the following quantity must be subtracted from the right-hand side of equation (3.39) to determine the active reserve:

$$\bar{P}_x \int_0^{m-t} {}_u p^{aa}_{x+t} v^u \, du. \tag{3.41}$$

The disabled reserve (see equation (3.40)), on the contrary, requires the subtraction of the following quantity, if $t < m$:

$$\bar{P}_x \int_0^{m-t} {}_u p^{ia}_{x+t} v^u \, du. \tag{3.42}$$

Now assume $\Gamma = [0, n, 0, \infty, r]$, with $r > n$. The active reserve is defined for $0 \le t \le n$. In the case of a single premium, its expression is given by (see equation (3.37)):

$$\bar{V}^a_{x+t,\Gamma} = \bar{a}^{ai}_{x+t,\Gamma} = \int_0^{n-t} {}_u p^{aa}_{x+t} \mu^{ai}_{x+t+u} v^u \bar{a}^{ii}_{x+t+u:\overline{r-t-u}|} \, du. \tag{3.43}$$

The formula for the case of continuous (constant) premium is straightforward: the quantity given by expression (3.41) must be subtracted from the right-hand side of equation (3.43) if $t < m$.

The disabled reserve is defined for $0 < t \le r$. In the case of a single premium we have:

$$\bar{V}^i_{x+t,\Gamma} = \int_0^{n-t} {}_u p^{ii}_{x+t} v^u \, du + {}_{n-t} p^{ii}_{x+t} \int_0^{r-n} {}_u p^{ii}_{x+n} v^u \, du \quad 0 < t \le n \tag{3.44}$$

$$\bar{V}^i_{x+t,\Gamma} = \int_0^{r-t} {}_u p^{ii}_{x+t} v^u \, du = \bar{a}^{ii}_{x+t:\overline{r-t}|} \quad n < t \le r \tag{3.44'}$$

Note that the first term of the right-hand side of equation (3.44) includes the expected present value of benefits payable up to time n, relating to the current disability spell as well as to possible future disability spells commencing before the term n. Conversely, the second term of the right-hand side of equation (3.44) is the expected present value of the benefits payable after time n relating to the last disability spell (commenced before time n).

Formula (3.44') expresses the fact that the disabled reserve after time n concerns only the current disability spell and that no further disability spell, commencing after the possible recovery, is taken into account.

In the case of continuous premium, if $t < m$ quantity (3.42) must be subtracted from the right-hand side of equation (3.44).

As far as single recurrent premiums are concerned, reserving problems will be dealt with in section 3.3.5.

3.2.7 Probabilities for the time-discrete models

A time-discrete Markov model for disability benefits has already been presented in Chapter 2, sections 2.1, 2.2 and 2.3. So, we can now restrict our attention to some notational aspects and some approximation formulae used in practice.

Let us consider the three-state disability model of Example 9, Chapter 1, in which recovery is possible. In a time-discrete context, we define the following probabilities related to an active insured aged y (as usual, we assume that no more than one transition can occur during one year, apart from the possible death of the insured).

p_y^{aa} = probability of being active at age $y + 1$;

q_y^{aa} = probability of dying within one year, the death occurring in state a;

p_y^{ai} = probability of being disabled at age $y + 1$;

q_y^{ai} = probability of dying within one year, the death occurring in state i;

p_y^{a} = probability of being alive at age $y + 1$;

q_y^{a} = probability of dying within one year;

w_y = probability of becoming disabled within one year.

The following relations hold:

$$p_y^{a} = p_y^{aa} + p_y^{ai} \tag{3.45}$$

$$q_y^{a} = q_y^{aa} + q_y^{ai} \tag{3.46}$$

$$p_y^{a} + q_y^{a} = 1 \tag{3.47}$$

$$w_y = p_y^{ai} + q_y^{ai}. \tag{3.48}$$

According to the notation used in Chapters 1 and 2 (and assuming now that the insured is aged y at time 0) we have:

$$p_y^{aa} = P_{11}(0,1) = P_{11}[0]$$

$$p_y^{ai} = P_{12}(0,1) = P_{12}[0]$$

$$q_y^a = P_{13}(0,1) = P_{13}[0].$$

Note that q_y^{aa} and q_y^{ai}, as well as w_y, are not defined for the Markov chain which is based on a one-year scanning of time. For this reason, in some actuarial literature the symbols q_y^{aa} and q_y^{ai} are not used, whereas q_y^a is replaced by p_y^{ad} which is consistent with the Markov state-oriented notation (an example is given in section 3.4.2).

Now, let us consider a disabled insured aged y. We define the following probabilities.

p_y^{ii} = probability of being disabled at age $y + 1$;

q_y^{ii} = probability of dying within one year, the death occurring in state i;

p_y^{ia} = probability of being active at age $y + 1$;

q_y^{ia} = probability of dying within one year, the death occurring in state a;

p_y^i = probability of being alive at age $y + 1$;

q_y^i = probability of dying within one year;

r_y = probability of recovery within one year.

The following relations hold:

$$p_y^i = p_y^{ia} + p_y^{ii} \tag{3.49}$$

$$q_y^i = q_y^{ii} + q_y^{ia} \tag{3.50}$$

$$p_y^i + q_y^i = 1 \tag{3.51}$$

$$r_y = p_y^{ia} + q_y^{ia}. \tag{3.52}$$

The symbol p_y^{id} often replaces q_y^i (see, for example, section 3.4.2). Conversely, note that q_y^{ii} and q_y^{ia} cannot be defined for a time-discrete Markov chain.

If we assume that no more than one transition can occur during one year (apart from possible death), probabilities p_y^{aa} and p_y^{ii} actually represent probabilities of remaining active and disabled respectively.

The set of probabilities needed for actuarial calculations can be reduced by adopting some approximation formulae. For example, we can assume:

$$q_y^{ai} = w_y \frac{q_y^{ii}}{2};$$ (3.53)

$$q_y^{ia} = r_y \frac{q_y^{aa}}{2}.$$ (3.54)

The probabilities above mentioned refer to a one-year period. Of course, we can define probabilities relating to two or more years. The notation is as in the time-continuous case. So we have, for example: $_h p_y^{aa}$, $_h p_y^{ai}$, etc. Note that (having excluded the possibility of more than one transition throughout the year) the occupancy probabilities can be expressed by the following relations:

$$_h p_y^{\overline{aa}} = \prod_{k=0}^{h-1} p_{y+k}^{aa}$$ (3.55)

$$_h p_y^{\overline{ii}} = \prod_{k=0}^{h-1} p_{y+k}^{ii}.$$ (3.56)

When actuarial values refer to the disability state, inception-select probabilities are often used. For example, $p_{[y]}^{ii}$ denotes the probability of being disabled at age $y + 1$, conditional on being disabled at age y with a duration of disability less than one year. The inception select occupancy probability is then given by:

$$_h p_{\overline{[y]}}^{ii} = \prod_{k=0}^{h-1} p_{[y]+k}^{ii}.$$ (3.57)

3.2.8 Allowing for lapses

The model represented by Fig. 3.1 could be extended by the inclusion of a fourth state to represent policies that have lapsed or been withdrawn (or terminated), denoted by w (see Fig. 3.5). It is unlikely that lives in state i would lapse their policies, so there would be transitions from a to w (and from w to d).

The inclusion of lapses in the model would be an important practical development because it is likely that lives in the active state are heterogeneous with respect to their propensity to become invalid or disabled. If proportionately fewer of the unhealthy policies withdraw by lapse (or surrender) and if these withdrawals occur predominantly in the early years after the issue of policy, the result would be a form of 'issue selection' that could persist into the higher durations of the policy. In the literature, this phenomenon has become known as 'selective withdrawals' and would be a feature of many types of insurance of the

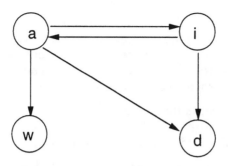

Figure 3.5 Multistate model allowing for lapses.

person, including life insurance, critical illness insurance and LTC insurance, as well as disability insurance. In some applications, the data available may not permit the distinguishing of terminations from deaths, and, as an expedient, it may be necessary to combine the two states.

3.2.9 Heterogeneity

The explicit modelling of 'heterogeneity' can be taken forward by the inclusion of 'frailty', or susceptibility to invalidity or disablement as a covariate in the transition intensity from the state a to i. For example, assuming that frailty is constant for each individual, we could consider the intensity $_z\mu_y^{ai}$ where z represents the individual frailty. A simple modelling assumption would then be:

$$_z\mu_y^{ai} = z \cdot {_1\mu_y^{ai}} \tag{3.58}$$

where an individual with a frailty of 1 is considered, in some sense, 'standard'.

3.2.10 Allowing for degree of disability

The model of Fig. 3.1 can also be extended by the replacement of state i by two states, representing disability of differing severity. This would be an appropriate model for contracts which paid a differential level of annuity, depending on the severity of disability – for example, a higher benefit in the acute phase and a lower benefit in respect of any residual disability. Such models will be discussed in more detail in Chapter 6 in the context of long-term care insurance (for example, see Fig. 6.2). Moreover, a model for 'graded disability' annuities will be presented in section 3.3 in the context of the Norwegian method.

3.2.11 IBNR reserves

A technical reserving problem arises in respect of group disability insurance (e.g. PHI in the UK) written on, say, a one-year basis. Consider business with a deferred period of f. At the end of the one-year risk period, the insurer will need to set up reserves in respect of:

1. lives who are currently claiming;
2. lives who are currently sick but whose period of sickness has not yet reached the deferred period f and so are not currently claiming.

The reserve for category 2 is an IBNR (incurred but not reported)-type reserve, widely discussed in non-life insurance. For a discussion of the formulae needed to estimate such reserves, readers are referred to Waters (1989).

3.3 SOME PARTICULAR CALCULATION PROCEDURES FOR DISABILITY ANNUITIES – I

In this section and in section 3.4 we discuss, in the light of the general model presented in the previous section, some calculation techniques that have been implemented in the literature and in actuarial practice. In particular we will see that (as pointed out by Waters, 1989):

- simpler models can be derived from a more complex model;
- a more complex model can be used to check the approximations and simplifications inherent in a simpler model.

In particular, this section deals with calculation procedures which naturally arise from the time-continuous Markov assumption, or can be rather easily interpreted within this context. Conversely, section 3.4 deals with time-discrete calculation procedures and with some traditional methods, including approximation formulae.

3.3.1 The Danish model

Let us consider a policy with $\Gamma = [0, n, 0, \infty, n]$. Thus, the insured period is $[0, n]$, with zero waiting period, zero deferred period, no maximum number of years of benefits operating, the stopping time equal to the policy term. Let us suppose that the probability model does not allow for recoveries. Hence, the following transition intensities shall be considered: μ_y^{ai}, μ_y^{ad}, μ_y^{id}. Danish insurance companies assume in their calculations that the transition intensities are given by:

$$\mu_y^{ai} = 0.0004 + 10^{0.06y - 5.46}$$

$$\mu_y^{ad} = \mu_y^{id} = 0.0005 + 10^{0.038y - 4.12} = \mu_y.$$

Thus, the (non-select) mortality of disabled lives is assumed to be equal to the mortality of active lives. The transition intensities assumed by the model follow the Makeham law. Note that from $\mu_y^{ad} = \mu_y^{id}$ it follows that active and disabled lives have the same survival probabilities, i.e.

$$_tp_y^{aa} + {}_tp_y^{ai} = {}_tp_y^{ii} = {}_tp_y. \tag{3.59}$$

Hence, we have:

$$_tp_y = \exp\left[-\int_0^t \mu_{y+u}\, du\right] \tag{3.60}$$

$$_tp_y^{aa} = {}_tp_{\overline{y}}^{aa} = \exp\left[-\int_0^t (\mu_{y+u}^{ai} + \mu_{y+u}^{ad})\, du\right]. \tag{3.61}$$

The following actuarial values are requested in premiums and reserve calculations:

$$\bar{a}_{x:\overline{n}|}^{ai} = \int_0^n {}_tp_x^{ai}v^t\, dt \tag{3.62}$$

$$\bar{a}_{x:\overline{n}|}^{ii} = \int_0^n {}_tp_x^{ii}v^t\, dt = \int_0^n {}_tp_x v^t\, dt = \bar{a}_{x:\overline{n}|} \tag{3.63}$$

$$\bar{a}_{x:\overline{n}|}^{aa} = \int_0^n {}_tp_x^{aa}v^t\, dt = \bar{a}_{x:\overline{n}|} - \bar{a}_{x:\overline{n}|}^{ai} \tag{3.64}$$

where $\bar{a}_{x:\overline{n}|}$ denotes the actuarial value of an ordinary life annuity.

Formulae for premiums and reserves can be easily obtained from the general structure described in section 3.2. For convenience, we now derive the particular formulae for the Danish model.

In order to calculate the level premium, $\bar{P}_{x:m,n|}$, let us assume that premiums are waived during disability and that premium payments cease after m years, $m < n$, in order to avoid a negative active reserve close to maturity (this problem will be discussed in section 3.3.7).

Then, we have:

$$\bar{P}_{x:m,n|} = \frac{\bar{a}_{x:\overline{n}|}^{ai}}{\bar{a}_{x:\overline{m}|}^{aa}}. \tag{3.65}$$

The active reserve, $\bar{V}_{x+t,n-t}^a$, is given for $0 \le t < m$, by:

$$\bar{V}_{x+t,n-t}^a = \bar{a}_{x+t:\overline{n-t}|}^{ai} - \bar{P}_{x:m,n|}\bar{a}_{x+t:\overline{m-t}|}^{aa} \tag{3.66}$$

and, for $m \le t \le n$, by:

$$\bar{V}_{x+t,n-t}^a = \bar{a}_{x+t:\overline{n-t}|}^{ai} \tag{3.67}$$

while the disabled reserve, $\bar{V}_{x+t,n-t}^i$, is given by:

$$\bar{V}_{x+t,n-t}^i = \bar{a}_{x+t:\overline{n-t}|}^{ii} \tag{3.68}$$

(since a non-select model is adopted and no possibility of recovery is taken into consideration).

3.3.2 The CMIB model

The Continuous Mortality Investigation Bureau (CMIB) was set up in 1923 by the Institute and Faculty of Actuaries in the United Kingdom. It includes several sub-committees. The PHI sub-committee was set up in 1970 to investigate the experience of the permanent health insurance business. The traditional actuarial methodology for PHI business in the United Kingdom is known as the 'Manchester Unity' method (see section 3.4.3). This method deals with disability data by looking at the proportion of lives disabled in a given period (hence, it is based on 'prevalence' rates). In CMIB Report No. 12, a new methodology has been proposed, based on a multiple state disability model.

The model includes three states, namely 'active', 'disabled', 'dead', and four transitions, namely 'death of an active life', 'death of a disabled life', 'disablement' and 'recovery'. Therefore, the model is represented by the graph of Fig. 3.1. The probabilistic structure is defined by the related transition intensities. It is assumed that the intensities of transition from the 'active' state depend only on the attained age. On the contrary, the intensities of transition from the 'disabled' state are assumed to depend only on the attained age and the duration since disability onset. No account is taken of the past sickness story (for instance, the number of previous disability periods). Hence, a semi-Markov model is adopted for modelling disability and (according to our notation) the following intensity functions are involved:

$$\mu_y^{ad}, \quad \mu_y^{ai}, \quad \mu_{y,r}^{ia}, \quad \mu_{y,r}^{id} \tag{3.69}$$

where y denotes the current age and r denotes the duration of the current spell of disability.

Data relating to PHI policies permit the graduation of three intensities out of the set (3.69), namely $\mu_y^{ai}, \mu_{y,r}^{ia}, \mu_{y,r}^{id}$. Graduated values for the intensity μ_y^{ad} are taken to be those based on data sets for insured lives with similar characteristics.

Graduation problems associated with the CMIB multiple state model for PHI will be analysed in Chapter 4.

3.3.3 The Swedish model

Let us consider a policy with $\Gamma = [0, n, f, \infty, n]$. Thus, the insured period is $[0, n]$, with zero waiting period, deferred period equal to f, no maximum number of years of benefits operating, the stopping time equal to the policy term. The following formula for the single premium can be derived:

$$\bar{a}_{x,\Gamma}^{ai} = \int_0^{n-f} {}_u p_x^{aa}\, \mu_{x+u}^{ai} \cdot {}_f p_{x+u}^{\overset{ii}{}} v^{u+f} \left[\int_{u+f}^{n} {}_{t-u-f} p_{x+u+f}^{\overset{ii}{}} v^{t-u-f}\, dt \right] du. \quad (3.70)$$

Formula (3.70) can be easily interpreted by direct reasoning. Note that ${}_f p_{x+u}^{\overset{ii}{}}$ is the probability that the disability lasts as long as required by the deferment condition.

Modifying equation (3.70), let us assume a single premium, $\bar{a}_{x:n(f)]}^{ai}$, given by:

$$\bar{a}_{x:n(f)]}^{ai} = \int_0^{n-f} {}_u p_x \nu_{x+u} \cdot {}_f p_{x+u,0}^{\overset{ii}{}} v^{u+f} \left[\int_{u+f}^{n} {}_{t-u-f} p_{x+u+f,f}^{\overset{ii}{}} v^{t-u-f}\, dt \right] du.$$

$$(3.71)$$

Note that:

- the (correct) probability ${}_u p_x^{aa}$ is replaced by the survival probability ${}_u p_x = l_{x+u}/l_x$; however, this probability is not meaningful within the multistate Markov model, since it disregards the state at entry, i.e. at age x (an exception is given by the case in which it is assumed that $\mu_y^{ad} = \mu_y^{id}$; see section 3.3.1);
- the transition intensity μ_{x+u}^{ai} is replaced by the function ν_{x+u}, which we shall discuss below;
- as far as transitions from the disability state are concerned, the probabilistic model assumed is inception-select; hence, the $p^{\overset{ii}{}}$'s depend on the duration of the disability.

Finally, let us define

$$\lambda(y,z) = {}_z p_{y,0}^{\overset{ii}{}} \quad (3.72)$$

$$\nu(y,z) = \nu_y \cdot {}_z p_{y,0}^{\overset{ii}{}} = \nu_y \lambda(y,z); \quad (3.73)$$

thus, $\lambda(y,z)$ is a select survival function concerning the continuance of disablement, while $\nu(y,z)\,dy$ should be interpreted as the probability that a healthy insured aged y becomes disabled between age y and $y+dy$ and remains disabled for a time z at least (more information about the exact meaning of ν_y, and hence $\nu(y,z)$, shall be provided below). Then, using

$$_f p_{x+u,0}^{\overset{ii}{}} \cdot {}_{t-u-f} p_{x+u+f,f}^{\overset{ii}{}} = {}_{t-u} p_{x+u,0}^{\overset{ii}{}} \quad (3.74)$$

we obtain

$$\bar{a}_{x:n(f)]}^{ai} = \int_0^{n-f} \frac{l_{x+u}}{l_x} \nu(x+u,f) v^{u+f}$$

$$\times \left[\int_{u+f}^{n} \frac{\lambda(x+u,t-u)}{\lambda(x+u,f)} v^{t-u-f}\, dt \right] du. \quad (3.75)$$

In Swedish actuarial practice it is usual to estimate functions λ and ν, and this justifies equation (3.75) as an expression for the single premium. It should be noted that the Swedish method is based on the probability of becoming sick and remaining sick.

As premiums are waived during disability, the following formula should be used to calculate the annual premium to be paid until age $x + m$:

$$\bar{P}_{x:m,n\rceil} = \frac{\bar{a}^{ai}_{x:n(f)\rceil}}{\bar{a}^{aa}_{x:m\rceil}} \tag{3.76}$$

where $m \leq n - f$. As probabilities $_tp^{aa}_x$ are not considered in the context of the Swedish model, the actuarial value $\bar{a}^{aa}_{x:m\rceil}$ of the 'active annuity' cannot be calculated, and hence the following approximation is used:

$$\bar{a}^{aa}_{x:m\rceil} + \bar{a}^{ai}_{x:m\rceil} \cong \bar{a}_{x:m\rceil} \tag{3.77}$$

where $\bar{a}_{x:m\rceil}$ denotes the actuarial value of an ordinary life annuity (in which the state at age x is not considered). Moreover, if premiums are waived after a waiting period which we suppose equal to the waiting period, f, of the disability annuity, then the level premium is assumed to be (approximately) given by:

$$\bar{P}_{x:m,n\rceil} = \frac{\bar{a}^{ai}_{x:n(f)\rceil}}{\bar{a}_{x:m\rceil} - \bar{a}^{ai}_{x:m(f)\rceil}}. \tag{3.78}$$

The active reserve, $\bar{V}^a_{x+t,n-t}$, is given, for $0 \leq t < m$, by:

$$\bar{V}^a_{x+t,n-t} = \bar{a}^{ai}_{x+t:n-t(f)\rceil} - \bar{P}_{x:m,n\rceil}(\bar{a}_{x+t:m-t\rceil} - \bar{a}^{ai}_{x+t:m-t(f)\rceil}) \tag{3.79}$$

and, for $m \leq t \leq n - f$, by:

$$\bar{V}^a_{x+t,n-t} = \bar{a}^{ai}_{x+t:n-t(f)\rceil} \tag{3.80}$$

while the disabled reserve (when the waiting period has elapsed) is assumed to be equal to:

$$\bar{V}^i_{x+t,n-t,[y]} = \bar{a}^{ii}_{[y]+x+t-y:n-t\rceil} = \int_0^{n-t} {}_zp^{ii}_{[y]+x+t-y}v^z \, \mathrm{d}z \tag{3.81}$$

where y denotes the age at disability inception. Note that formula (3.81) does not allow for recovery and possible future spells of disability.

Let us now briefly illustrate the technical bases used to express mortality and disability. First, the quantity ν_y should not be regarded as an intensity of transition from the active state into the disability state. Actually, the function ν_y is estimated with reference to a population including healthy and disabled lives. Hence, the quantity

$$\nu(y, z) \, \mathrm{d}y = \nu_y \lambda(y, z) \, \mathrm{d}y$$

can be used to evaluate correctly the number of cases of disability, with a duration of z at least, which commenced in the age interval $[y, y + dy)$, in an insurance portfolio of active as well as disabled individuals. Note that this interpretation (partially) justifies the use of the survival probability $_up_x$ instead of the probability $_up_x^{aa}$.

The experience coming from portfolios with different waiting periods suggests the introduction of functions ν which depend, to some extent, on the waiting period f. For the sake of simplicity, the functional dependence has been assumed proportional, by using the following formula:

$$\nu_y^{(f)} = r(f)\nu_y \tag{3.82}$$

where $r(f)$ is independent of y.

An example of a technical basis is given by the following set of functions (which define the basis G73 for males):

$$\nu_y = \frac{0.4}{l_y}$$

$$l_y = e^{\alpha/\beta} \, e^{-\gamma y} \, e^{-(\alpha/\beta)e^{\beta y}}$$

with $\alpha = 0.00034, \beta = 0.042 \ln(10), \gamma = 0.00006$

$$\lambda(y, t) = a_y \, e^{-80t} + b_y \, e^{-13t} + c_y \, e^{-1.5t} + d_y(0.15 \, e^{-0.3t} + 0.85 \, e^{-0.04t})$$

with $\quad a_y = 1 - b_y - c_y - d_y, \quad b_y = 0.12, \quad c_y = 0.006 \, e^{0.04y}, \quad d_y = 0.001 + 0.000011 \, e^{0.13y}$

$$r(f) = \begin{cases} 2.3 - 10.8f & \text{if } 0 \le f < 1/12 \\ 1.6 - 2.4f & \text{if } 1/12 \le f < 1/4 \\ 1 & \text{if } f \ge 1/4. \end{cases}$$

Note that $\lambda(y, t)$ is a mixture of exponentials. The weights depend only on the age at disability inception, whilst the exponentials are functions of the disability duration. The terms relate to cases of presumably different expected durations.

3.3.4 The Norwegian model

Let us consider a policy defined by the set of conditions $\Gamma = [0, n, 0, \infty, n]$. Then, the insured period is $[0, n]$, with zero waiting period, zero deferred period, no maximum number of years of benefits operating, the stopping time equal to the policy term. Hence, the net single premium is given (according to the usual actuarial notation) by

$$\bar{a}_{x:n|}^{ai} = \int_0^n {}_tp_x^{ai}v^t \, dt. \tag{3.83}$$

Let $j_{(x)+t}$ denote the probability that an individual is disabled at age $x + t$ given that he was healthy at age x (age at policy issue) and that he is alive at age $x + t$:

$$j_{(x)+t} = \Pr\{S(x+t) = i \mid [(S(x) = a) \wedge [(S(x+t) = a) \vee (S(x+t) = i)]]\}$$

(3.84)

(according to the usual actuarial notation, $j_{(x)+t}$ denotes a function of the two variables x and t).

Since

$$\Pr\{(S(x+t) = a) \vee (S(x+t) = i) \mid S(x) = a\} = {}_tp_x^{aa} + {}_tp_x^{ai}$$

(3.85)

we immediately obtain

$${}_tp_x^{ai} = ({}_tp_x^{aa} + {}_tp_x^{ai})j_{(x)+t}$$

(3.86)

so that the single premium is given by

$$\bar{a}_{x:n]}^{ai} = \int_0^n ({}_tp_x^{aa} + {}_tp_x^{ai})j_{(x)+t}v^t \, dt.$$

(3.87)

Since the probability ${}_tp_x^{ai}$ is still involved, formula (3.87) is more 'complicated' than formula (3.83). If we approximate the probability $({}_tp_x^{aa} + {}_tp_x^{ai})$ by using a 'simple' survival probability ${}_tp_x = l_{x+t}/l_x$ (independent of the state at age x), we get the formula used in the so-called 'Norwegian method':

$$\bar{a}_{(x):n]}^{ai} = \int_0^n {}_tp_x j_{(x)+t}v^t \, dt.$$

(3.88)

It follows that the Norwegian method is based on the probability of being disabled. However, it should be stressed that the probability ${}_tp_x$ is not meaningful within the multistate Markov model, since it disregards the state at entry, i.e. at age x (see also section 3.3.3). Hence, the value of the integral at right-hand side of equation (3.88) is not a function well-defined in the context of this model; then, the actuarial value $\bar{a}_{(x):n]}^{ai}$ can be accepted only as a numerical approximation. Note that probabilities ${}_tp_x$ are simply based on a survival table, whilst the estimation of the probabilities ${}_tp_x^{ai}$ (used in the correct formula (3.83)) is not a trivial matter (unless the transition intensities are assigned; a particular case has been described in section 3.3.1).

Level premiums can be evaluated as follows. Let us suppose that premiums are paid while the insured is active, in the time interval $[0, m)$, $m \leq n$. Hence, an 'active annuity' shall be considered, the actuarial value of which is:

$$\bar{a}_{x:m]}^{aa} = \int_0^m {}_tp_x^{aa} v^t \, dt.$$

(3.89)

Then, we have:

$$\bar{P}_{x:m,n\rceil} = \frac{\bar{a}^{ai}_{x:n\rceil}}{\bar{a}^{aa}_{x:m\rceil}}. \tag{3.90}$$

In practice, the evaluation $\bar{a}^{aa}_{x:m\rceil}$ is not a trivial matter, since the transition intensities are not explicitly considered. To this purpose, note that

$$\bar{a}^{aa}_{x:m\rceil} + \bar{a}^{ai}_{x:m\rceil} = \int_0^m ({}_tp^{aa}_x + {}_tp^{ai}_x)v^t \, dt \tag{3.91}$$

is the actuarial value of an annuity payable by (or to) an insured, active or disabled, who is active at age x. If we accept the approximation which consists in replacing ${}_tp^{aa}_x + {}_tp^{ai}_x$ by ${}_tp_x = l_{x+t}/l_x$ (see formulae (3.87) and (3.88)), then, replacing $\bar{a}^{ai}_{x:m\rceil}$ with $\bar{a}^{ai}_{(x):m\rceil}$ (given by formula (3.88)), we obtain:

$$\bar{a}^{aa}_{x:m\rceil} = \int_0^m ({}_tp^{aa}_x + {}_tp^{ai}_x)v^t \, dt - \bar{a}^{ai}_{x:m\rceil} \cong \bar{a}_{x:m\rceil} - \bar{a}^{ai}_{(x):m\rceil} \tag{3.92}$$

where $\bar{a}_{x:m\rceil}$ denotes the actuarial value of a temporary life annuity. Finally, if it is assumed that

$$\bar{a}^{aa}_{(x):m\rceil} = \bar{a}_{x:m\rceil} - \bar{a}^{ai}_{(x):m\rceil} \tag{3.93}$$

we obtain the (approximate) expression for the level premium in the Norwegian method:

$$\bar{P}_{(x):m,n\rceil} = \frac{\bar{a}^{ai}_{(x):n\rceil}}{\bar{a}^{aa}_{(x):m\rceil}}. \tag{3.94}$$

Formulae from the Norwegian method can be applied also to disability covers in which the amount paid by the insurer depends on the degree of disability (thus, a 'graded disability annuity' is involved). For the sake of simplicity, let us assume that the amounts paid are proportional to the degree of disability. Let $G_{x,t}$ denote the random degree of disability for an insured who is disabled at age $x + t$, being active at age x; then $0 < G_{x,t} \le 1$. Let $E(G_{x,t})$ denote the expected value of $G_{x,t}$. Hence, the actuarial value of the disability annuity is given by:

$$\bar{a}^{ai}_{x:n\rceil} = \int_0^n {}_tp^{ai}_x E(G_{x,t})v^t \, dt. \tag{3.95}$$

According to the Norwegian approach, the (exact) single premium is given by formula (3.87). Equating the integrands in (3.87) and (3.95), we obtain:

$$j_{(x)+t} = {}_tp^{ai}_x \frac{E(G_{x,t})}{{}_tp^{aa}_x + {}_tp^{ai}_x}. \tag{3.96}$$

Then, the function $j_{(x)+t}$ can be interpreted as the expected degree of disability at age $x + t$ of the insured, given that he is active at age x and alive at age $x + t$.

Now, let us turn to mathematical reserves. Let us assume $m = n$. According to the Norwegian method, the active reserve, $\bar{V}^a_{(x)+t,n-t}$, is then given by:

$$\bar{V}^a_{(x)+t,n-t} = \bar{a}^{ai}_{(x+t):n-t\rceil} - \bar{P}_{(x):n,n\rceil}\bar{a}^{aa}_{(x+t):n-t\rceil}. \qquad (3.97)$$

Note that, writing $(x + t)$ according to the Norwegian notation, we mean that the insured is assumed to be active at age $x + t$.

The 'joint' reserve (i.e. the reserve which does not take into account the state of the insured at age $x + t$), is given by:

$$\bar{V}_{(x)+t,n-t} = \bar{a}^{ai}_{(x)+t:n-t\rceil} - \bar{P}_{(x):n,n\rceil}\bar{a}^{aa}_{(x)+t:n-t\rceil}. \qquad (3.98)$$

Note that $(x) + t$ indicates that the insured is now assumed to be active at age x and simply alive at age $x + t$.

In Norwegian actuarial practice, all of the insureds, i.e. active as well as disabled, are considered to form one single group as far as reserving is concerned. Thus, formula (3.98) is used for reserving. It follows that the portfolio reserve is independent of the actual number of disabled insureds at the time of the reserve calculation. The active reserve formula, on the other hand, is used to calculate paid-up values and surrender values for active lives.

An evaluation of the mathematical reserve for disabled lives, $\bar{V}^i_{(x)+t,n-t}$, can be obtained as follows. As $j_{(x)+t}$ is the probability that an individual is disabled at age $x + t$ given that he was healthy at age x and that he is alive at age $x + t$, we can think of the joint reserve as being a weighted average of the active reserve and the disabled reserve, the weights being respectively given by $1 - j_{(x)+t}$ and $j_{(x)+t}$. Hence:

$$\bar{V}_{(x)+t,n-t} = (1 - j_{(x)+t})\bar{V}^a_{(x)+t,n-t} + j_{(x)+t}\bar{V}^i_{(x)+t,n-t} \qquad (3.99)$$

from which $\bar{V}^i_{(x)+t,n-t}$ can be immediately calculated. Note, however, that the reserve $\bar{V}^i_{(x)+t,n-t}$ is independent of the duration of the disability. It follows that a method based on the (inception-select) actuarial value of a disability annuity would be preferable. It must be stressed that use of the multiple state model without approximation would give this desired result.

3.3.5 A model based on 'single recurrent premiums'

Group disability insurance is usually financed by a 'single recurrent premium' mechanism, which represents a particular discrete-time premium arrangement. As far as disability benefits are concerned, a single recurrent premium is usually calculated as a natural premium. By definition, the natural premium pertaining to an insured aged x (at the beginning of the contract year) must meet the expected cost of a disability annuity which might commence between age x and $x + 1$. The related actuarial model represents a particular case of the actuarial model describing a disability annuity risk where benefits are paid only if a claim occurs during a restricted period (one year, in the natural premium model), but where benefit payments may extend beyond this period once they have begun. The model we now describe has been proposed by Hoem (1988).

First let us describe the natural premium model in terms of insurance cover conditions. Assume that the disability annuity is payable until retirement age ξ. Let $r = \xi - x$, where x indicates the age at the beginning of the concerned year. For simplicity, assume that there is no deferred period. Then, the set of cover conditions is $\Gamma = [0, 1, 0, \infty, r]$. Thus, the insured period is $[0, 1]$, with zero waiting period, zero deferred period, no maximum number of years of benefit operating, the stopping time equal to r. This model represents a particular case of the more general model defined by $\Gamma = [0, n, 0, \infty, r]$; hence, actuarial values can be evaluated using formulae presented in section 3.2 (see, for example, (3.31)).

The related ϕ^Γ function (see section 3.2.5, and formula (3.20) in particular) is as follows:

$$\phi^{[0,1,0,\infty,r]}(x, t) = \begin{cases} {}_t p_x^{ai} & \text{if } t < 1 \\ {}_t p_x^{ai}(t - 1) & \text{if } 1 \le t < r \\ 0 & \text{if } t \ge r. \end{cases} \tag{3.100}$$

Let $\bar{a}_{x,\Gamma}^{ai}$ denote the actuarial value of the benefits related to claims which may occur in the age interval $(x, x + 1)$; this is given by:

$$\bar{a}_{x,\Gamma}^{ai} = \int_0^{+\infty} \phi^{[0,1,0,\infty,r]}(x, t) v^t \, dt$$

$$= \int_0^1 {}_t p_x^{ai} v^t \, dt + \int_1^r {}_t p_x^{ai}(t - 1) v^t \, dt. \tag{3.101}$$

Note that expression (3.101) can be directly obtained from equation (3.31) with $n = 1$. After some manipulations (according to the procedure described in section 3.2.6) we obtain:

$$\bar{a}_{x,\Gamma}^{ai} = \int_0^1 {}_u p_x^{aa} \mu_{x+u}^{ai} v^u \bar{a}_{x+u:\overline{r-u}|}^{ii} \, du \tag{3.102}$$

which represents a particular case of (3.37), namely with $n = 1$. It is self-evident that $\bar{a}_{x,\Gamma}^{ai}$ equals the expected costs of disability annuities which might commence between age x and $x + 1$. Note that, for simplicity, a non-select scheme has been adopted.

Now, let us shift to age $x + 1$: the end of the contract year. If the insured is disabled, a disability reserve must be assessed. This reserve is given by:

$$\bar{V}^i_{x+1} = \bar{a}^{ii}_{x+1:r-1|} \tag{3.103}$$

(which represents a particular case of (3.44')). If the insured is active, no reserve is required, as a natural premium mechanism works.

The problem we now wish to address is as follows. The whole group insurance policy is a single-year transaction that can be terminated at the end of any contract year without any obligation, apart from the disability annuities then running. However, renewal of a group policy at the end of each contract year is usually quite automatic. In this case, it is assumed that the cover will work for each individual until some fixed age, e.g. the retirement age. Hence, it is very important to realize that the sequence of single recurrent premiums (related to each insured) must meet the expected costs pertaining to claims which may occur until the retirement age. Unfortunately, this is not the case for the sequence of natural premiums $\bar{a}^{ai}_{x,\Gamma}$. Actually:

- if the insured aged $x+1$ is active, a premium $\bar{a}^{ai}_{x+1,\Gamma}$ is paid, meeting the costs related to claims in $(x+1, x+2)$;
- if the insured aged $x+1$ is disabled, no premium will be paid until the first anniversary following his recovery (provided that the recovery occurs before age $\xi - 1$).

It is easy to realize that the sequence of natural premiums does not necessarily meet the expected costs. Figure 3.6 illustrates two situations having different consequences in terms of meeting the expected costs. The disability story illustrated by Fig. 3.6a implies premium payment at age x as well as at age $x+2$. The second premium meets the expected cost related to the contract year $(x+2, x+3)$, and hence funds the

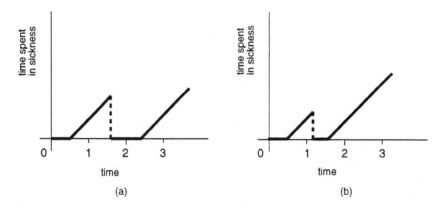

Figure 3.6 Disability stories with different consequences on natural premium payment.

claims occurring in that year. On the contrary, Fig. 3.6b represents a disability story which does not allow for completely meeting the costs: actually, the second disability spell (commencing between age $x + 1$ and $x + 2$) is unfunded.

Then, the premium mechanism must be modified. Two alternative solutions can be adopted.

1. Leaving unchanged the amounts of premiums paid when the insured is active, an additional premium is paid upon recovery. The amount of this addition can be assumed to be proportional to the time to next anniversary, so that the integration itself meets (approximately) the claims which may occur in the remaining part of the year.
2. A new approach can be adopted in calculating single recurrent premiums, to be paid at the policy anniversaries only, whatever the individual disability story might be. The calculation principle must take into account the disability spells which might start after a recovery and before the payment of a premium.

In the following, we will deal with approach 2.

Let $\bar{P}_{x,\Gamma}$ denote the single recurrent premium, to be determined according to approach 2. The relevant calculations require the following probability:

$$_t p_x^{iai} = \Pr\{(S(x+t) = i) \wedge (S(x+\tau) = a$$

$$\text{for some } \tau \in (0,t)) \mid S(x) = i\}. \tag{3.104}$$

Moreover, let

$$\bar{a}_{x:u\rceil}^{iai} = \int_0^u {}_t p_x^{iai} v^t \, dt \tag{3.105}$$

and, as usual:

$$\bar{a}_{x:u\rceil}^{ai} = \int_0^u {}_t p_x^{ai} v^t \, dt. \tag{3.106}$$

The actuarial value $\bar{a}_{x,\Gamma}^{ai}$, given by (3.102), can be also expressed as follows:

$$\bar{a}_{x,\Gamma}^{ai} = \bar{a}_{x:r\rceil}^{ai} - v({}_1 p_x^{ai} \bar{a}_{x+1:r-1\rceil}^{iai} + {}_1 p_x^{aa} \bar{a}_{x+1:r-1\rceil}^{ai}). \tag{3.107}$$

Formula (3.107) can be easily interpreted by direct reasoning: the positive term on the right-hand side collects all disability stories commencing between age x and $x + r = \xi$; the negative term includes all disability spells which commence between age $x + 1$ and $x + r$, respectively assuming that the insured is disabled or active at age $x + 1$.

The active reserve at age $x + t$ (which must be equal to zero at each anniversary, as a natural premium arrangement works) is given for

$0 \leq t \leq r$ by:

$$\bar{V}^a_{x+t} = \bar{a}^{ai}_{x+t:r-t\rceil} - \sum_{k=\lceil t\rceil}^{r-1} v^{k-t}{}_{k-t}p^{aa}_{x+t}\bar{P}_{x+k,\Gamma} \qquad (3.108)$$

where $\lceil t\rceil$ denotes the smallest integer greater than or equal to t. From the constraint

$$\bar{V}^a_{x+t} = 0 \quad \text{for } t = 0, 1, \ldots, r$$

the following conditions for the single recurrent premiums are immediately obtained:

$$\bar{P}_{x+r-1,\Gamma} = \bar{a}^{ai}_{x+r-1:1\rceil} \qquad (3.109)$$

$$\bar{P}_{x+k,\Gamma} = \bar{a}^{ai}_{x+k:r-k\rceil} - \sum_{h=1}^{r-k-1} v^h {}_h p^{aa}_{x+k}\bar{P}_{x+k+h,\Gamma}$$

$$\text{for } k = r-2, r-3, \ldots, 1, 0. \qquad (3.110)$$

Formula (3.109) is self-evident. Formula (3.110) can be easily explained. Each single recurrent premium must exactly meet the actuarial value of the disability annuities which are not funded by future premiums, provided that these premiums will be paid (i.e. the insured will be active).

The disability reserve must meet the running annuity as well as the expected costs relating to future disability spells, which may commence after recovery but before premium payment. Hence, we obtain:

$$\bar{V}^i_{x+t} = \bar{a}^{ii}_{x+t:r-t\rceil} + \bar{a}^{iai}_{x+t:r-t\rceil} - \sum_{k=\lceil t\rceil}^{r-1} v^{k-t}{}_{k-t}p^{ia}_{x+t}\bar{P}_{x+k,\Gamma}. \qquad (3.111)$$

Finally, it is interesting to compare the natural premiums $\bar{a}^{ai}_{x+k,\Gamma}$ with the single recurrent premiums $\bar{P}_{x+k,\Gamma}$. From formula (3.110) we obtain:

$$\bar{P}_{x+k,\Gamma} = \bar{a}^{ai}_{x+k:r-k\rceil} - v \cdot {}_1 p^{aa}_{x+k}\bar{P}_{x+k+1,\Gamma} - \sum_{h=2}^{r-k-1} v^h {}_h p^{aa}_{x+k}\bar{P}_{x+k+h,\Gamma}. \qquad (3.112)$$

Let us then express the premium $\bar{P}_{x+k+1,\Gamma}$ by using formula (3.110) and insert this expression into (3.112); after some manipulations, we obtain:

$$\bar{P}_{x+k,\Gamma} = \bar{a}^{ai}_{x+k,\Gamma} + {}_1 p^{ai}_{x+k}\left[v\bar{a}^{iai}_{x+k+1:r-k-1\rceil} - \sum_{h=2}^{r-k-1} v^h {}_{h-1}p^{ia}_{x+k+1}\bar{P}_{x+k+h,\Gamma} \right].$$

$$(3.113)$$

Hence, the premium $\bar{P}_{x+k,\Gamma}$ exactly meets the expected costs referring to:

- benefits related to disability spells commencing between age $x + k$ and $x + k + 1$;

- benefits related to disability spells following the first recovery after age $x + k + 1$, this anniversary belonging to a disability spell, but exclusively to the extent that these benefits are not met by future premiums.

3.3.6 The Finnish model

We shall illustrate some aspects of the disability model adopted by the Finnish pension schemes based on the Employee's Pension Act (TEL). In the mandatory part of TEL, there is no funding in advance of disability pensions, which are funded at the time of commencement by a natural premium mechanism. Hence, we shall mainly concentrate our discussion on actuarial aspects of disability pensions once they have begun.

Two features characterize the Finnish TEL disability model:

1. the choice of a particular function for describing the disability duration;
2. the choice of a particular model (i.e. a mixture of exponentials) for the above-mentioned function.

First, let us introduce the function $Z(y, u)$. The Finnish model expresses the probability, $P(y; u_1, u_2)$, for a person aged 0 to live to age y and to be disabled at age y with the past duration of disability belonging to the interval (u_1, u_2), as follows:

$$P(y; u_1, u_2) = \int_{u_1}^{u_2} Z(y, u) \, \mathrm{d}u. \tag{3.114}$$

In terms of the probabilities p^{ii}'s of remaining disabled, the following relation obviously holds:

$$Z(y + t, u + t) \, \mathrm{d}u = Z(y, u) \, _t p^{ii}_{y,u} \, \mathrm{d}u \tag{3.115}$$

so that (through a limit procedure) we obtain:

$$_t p^{ii}_{y,u} = \frac{Z(y + t, u + t)}{Z(y, u)}. \tag{3.116}$$

The (select) actuarial value of a disability annuity until age ξ (e.g. the retirement age) is given by:

$$\bar{a}^{ii}_{[y - u] + u : \overline{\xi - y}|} = \int_0^{\xi - y} {}_t p^{ii}_{y,u} v^t \, \mathrm{d}t = \int_0^{\xi - y} {}_t p^{ii}_{y,u} \, \mathrm{e}^{-\delta t} \, \mathrm{d}t \tag{3.117}$$

where y is the current age, u is the past duration of disability and δ denotes the force of interest. Using (3.116), we have:

$$\bar{a}^{ii}_{[y - u] + u : \overline{\xi - y}|} = \frac{1}{Z(y, u)} \int_0^{\xi - y} Z(y + t, u + t) \, \mathrm{e}^{-\delta t} \, \mathrm{d}t. \tag{3.118}$$

Some connections among various probabilities will be useful when dealing with premium calculations. In particular, let us assume that a newborn is, at least for a moment, an 'active' individual. Hence, the following relation holds:

$$Z(y, u)\, du = {}_{y-u}p_0^{aa}\, \mu_{y-u}^{ai} \cdot {}_u p_{y-u,0}^{ii}\, du. \tag{3.119}$$

Now, let us deal with some particular hypotheses adopted by the Finnish model. First, let us suppose that the (select) probability of remaining disabled can be expressed by a mixture of J negative exponentials, i.e.:

$$_t p_{y,u}^{ii} = \sum_{j=1}^{J} w_j(y, u)\, e^{-\mu_j t} \tag{3.120}$$

where y is the current age of the disabled and u is the past duration of the disability. Note that such a structure allows for the separate modelling of cases of disability having different expected durations. In the context of the Finnish model, the function $Z(y, u)$ is assumed to be a mixture of exponentials:

$$Z(y, u) = \sum_{j=1}^{3} a_j\, e^{b_j y - c_j u} = \sum_{j=1}^{3} Z_j(y, u) \tag{3.121}$$

whence:

$$_t p_{y,u}^{ii} = \sum_{j=1}^{3} \frac{a_j\, e^{b_j(y+t) - c_j(u+t)}}{Z(y, u)} = \sum_{j=1}^{3} \frac{Z_j(y, u)\, e^{(b_j - c_j)t}}{Z(y, u)} \tag{3.122}$$

which is a particular case of (3.120), namely with $J = 3$ and $\mu_j = -(b_j - c_j)$.

For the actuarial value of a disability annuity (i.e. the disabled reserve, when only permanent disability is considered), we obtain from equations (3.117) and (3.120):

$$\bar{a}_{[y-u]+u:\overline{\xi-y}|}^{ii} = \sum_{j=1}^{J} \left[w_j(y, u) \int_0^{\xi-y} e^{-(\delta+\mu_j)t}\, dt \right]. \tag{3.123}$$

Let us adopt the following notation for the present value of an annuity certain:

$$\bar{a}_{\overline{n}|}^{(\delta+\mu)} = \int_0^n e^{-(\delta+\mu)t}\, dt = \frac{1 - e^{-(\delta+\mu)n}}{\delta + \mu}. \tag{3.124}$$

Hence:

$$\bar{a}_{[y-u]+u:\overline{\xi-y}|}^{ii} = \sum_{j=1}^{J} w_j(y, u)\, \bar{a}_{\overline{\xi-y}|}^{(\delta+\mu_j)}. \tag{3.125}$$

Finally, let us illustrate the calculation of premiums. Insurance cover conditions usually include a deferred period, f, and a maximum number of years of annuity payment (from disability inception), m. For simplicity, we first consider the case in which $f = 0$ and $m = \infty$ (see section 3.2.2), thus $\Gamma = [0, r, 0, \infty, r]$ with $r = \xi - x$. Note that the longest period of disability annuity payment is implicitly given by $\xi - y$, when y is the age at disability inception.

Then, for the single premium, we have:

$$\bar{a}_{x,\Gamma}^{ai} = \int_0^r e^{-\delta t} \, {}_t p_x^{ai} \, dt = \int_0^r \int_0^t e^{-\delta t} \, {}_s p_x^{aa} \, \mu_{x+s}^{ai} \cdot {}_{t-s} p_{x+s,0}^{ii} \, ds \, dt. \tag{3.126}$$

Let us take ${}_s p_x$ as an approximation of ${}_s p_x^{aa}$. Since ${}_s p_x = {}_{x+s} p_0 / {}_x p_0$, we obtain:

$$\bar{a}_{x,\Gamma}^{ai} \cong \frac{1}{e^{-\delta x} \, {}_x p_0} \int_0^r \int_0^t e^{-\delta(x+t)} \, {}_{x+s} p_0 \mu_{x+s}^{ai} \cdot {}_{t-s} p_{x+s,0}^{ii} \, ds \, dt. \tag{3.127}$$

Replacing ${}_{x+s} p_0$ by ${}_{x+s} p_0^{aa}$ again, and defining $z = t - s$ (time spent in disability), we have:

$$\bar{a}_{x,\Gamma}^{ai} \cong \frac{1}{e^{-\delta x} \, {}_x p_0} \int_0^r \int_0^t e^{-\delta(x+t)} Z(x+t, z) \, dz \, dt. \tag{3.128}$$

Finally, adopting the hypothesis of a constant intensity of mortality (as the Finnish model does), we have:

$$\bar{a}_{x,\Gamma}^{ai} \cong \frac{1}{e^{-(\delta+\mu)x}} \int_0^r \int_0^t e^{-\delta(x+t)} Z(x+t, z) \, dz \, dt. \tag{3.129}$$

The right-hand side of equation (3.129) (approximately) expresses the single premium in terms of the function Z of the Finnish model. Note that the double integral involves an actuarial evaluation at age 0 (which is consistent with the meaning of the function Z), while the term $1/e^{-(\delta+\mu)x}$ shifts the evaluation to age x.

Let us now illustrate the annual natural premium calculation. The natural premium at age x must meet the expected cost of a disability annuity which might commence between age x and $x + 1$. Let us denote by $\bar{P}_x^{(N)}$ the natural premium. We find (see section 3.3.5 and, in particular, function (3.100)):

$$\bar{P}_x^{(N)} = \int_0^1 \int_s^r e^{-\delta t} \, {}_s p_x^{aa} \, \mu_{x+s}^{ai} \cdot {}_{t-s} p_{x+s,0}^{ii} \, dt \, ds. \tag{3.130}$$

Then, using the approximations adopted when calculating the single premium, we obtain:

$$\bar{P}_x^{(N)} \cong \frac{1}{e^{-(\delta+\mu)x}} \int_0^1 \int_s^r e^{-\delta(x+t)} {}_{x+s}p_0^{aa}\,\mu_{x+s}^{ai} \cdot {}_{t-s}p_{x+s,0}^{ii}\, dt\, ds$$

$$= \frac{1}{e^{-(\delta+\mu)x}} \int_0^1 \int_s^r e^{-\delta(x+t)} Z(x+t, t-s)\, dt\, ds \qquad (3.131)$$

which gives the risk premium in terms of the function Z. Approximating the integral $\int_0^1$ with the value of the integrand function at $s = \frac{1}{2}$, we finally obtain:

$$\bar{P}_x^{(N)} \cong \frac{1}{e^{-(\delta+\mu)x}} \int_{1/2}^r e^{-\delta(x+t)} Z(x+t, t-\tfrac{1}{2})\, dt. \qquad (3.132)$$

More realistic cover conditions lead to more complicated formulae. For example, a realistic set of conditions is defined by $\Gamma = [0, r, f, m, r]$ with $r = \xi - x$ and $m \le r$. In this case, the longest period of annuity payment is m. The corresponding single premium is (approximately) given by:

$$\bar{a}_{x,\Gamma}^{ai} \cong \frac{1}{e^{-(\delta+\mu)x}} \Bigg[-\int_f^r \int_0^f e^{-\delta(x+t)} Z(x+t, z)\, dz\, dt$$

$$+ \int_f^m \int_0^t e^{-\delta(x+t)} Z(x+t, z)\, dz\, dt$$

$$+ \int_m^r \int_0^m e^{-\delta(x+t)} Z(x+t, z)\, dz\, dt \Bigg]. \qquad (3.133)$$

Formula (3.133) can be justified by using the apparatus presented in section 3.2.5.

3.3.7 A numerical example

We now present a numerical example which illustrates premiums and reserves for a disability annuity benefit. The constant rate of the disability annuity is $b = 1$. Results are presented for convenience in graphical form.

Calculations have been performed assuming the transition intensities used by Danish insurance companies (see section 3.3.1). However, as far as the mortality of the disabled lives is concerned, we have assumed the hypothesis $\mu_y^{id} = (1+\eta)\mu_y^{ad}$ (i.e. a differential mortality has been assumed, with $\eta = 0.20$). Moreover, several values of the extra level of mortality have been considered, in order to perform a sensitivity analysis concerning the disabled reserve.

Premiums and reserves have been calculated using formulae presented in section 3.2, taking into account the fact that the Danish model does not allow for recovery. This involves some simplifications. For example we

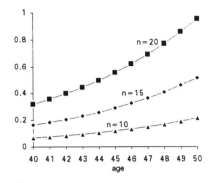

Figure 3.7 Single premium as a function of the age x at entry.

have $_up^{ii}_{x+t} = {_up^{\overline{ii}}_{x+t}}$, whence formula (3.44) simply becomes:

$$\bar{V}^i_{x+t,\Gamma} = \int_0^{r-t} {_up^{\overline{ii}}_{x+t}} v^u \, du \quad 0 < t \le n.$$

Firstly, let us consider a policy with $\Gamma = [0, n, 0, \infty, n]$. In Figs 3.7 and 3.8 the single premium as well as the continuous constant premium are represented as a function of the age x at entry, for various policy terms. The continuous premium is assumed to be payable for n years.

In Fig. 3.9 the active reserve is depicted for different sets of policy conditions and premium payment periods. Note that r denotes the stopping time (from policy issue) of the disability annuity payment, while m denotes the premium payment term. According to the notation used in section 3.2 we have $\Gamma = [0, n, 0, \infty, r]$ (see in particular Fig. 3.4h).

The arrangement $n = r = m = 20$ implies a negative active reserve, $\bar{V}^a_{x+t,m-t}$, throughout the last six policy years. This is due to the fact that the 'risk' is decreasing as t increases, because the possible duration of the disability annuity is decreasing, whilst the continuous premium

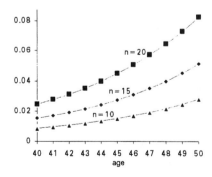

Figure 3.8 Continuous premium as a function of the age x at entry.

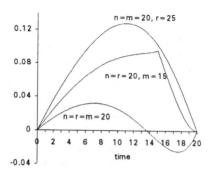

Figure 3.9 Active reserve in the case of continuous premium.

is assumed to be constant. In order to avoid a negative reserve close to maturity, premium payments can be limited to a shorter period, $m < n$ (as is advocated by the standard textbooks, for example Neill (1977), in the case of life insurance mathematics). The reserve relating to the case $n = r = 20$, $m = 15$, illustrates this solution. Note that when $n = m = 20$ but $r = 25$ the active reserve is positive throughout the policy duration.

Figure 3.10 illustrates the behaviour of the disabled reserve, $\bar{V}^i_{x+t, m-t}$, when $r = 20$ and $r = 25$. Since a non-inception-select scheme is adopted, the reserve only depends on the attained age (and hence on the duration since policy issue), whilst it is independent of the duration since disability inception.

In Fig. 3.11 the disabled reserves for some values of η are depicted, for $t \leq 3$. It is evident that the extra level of mortality has a very low effect on the disabled reserve.

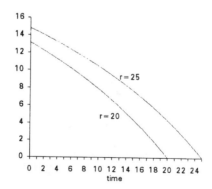

Figure 3.10 Disabled reserve.

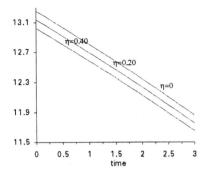

Figure 3.11 Disabled reserve as a function of the extra level of mortality (parameter η).

3.4 SOME PARTICULAR CALCULATION PROCEDURES FOR DISABILITY INSURANCE – II

3.4.1 The inception/annuity models

Let us return to equation (3.24), which can be rewritten in the following form:

$$\bar{a}_x^{ai} = \sum_{h=0}^{+\infty} \int_0^1 {}_{h+u}p_x^{aa}\, \mu_{x+h+u}^{ai}\, v^{h+u}$$

$$\times \left[\int_{h+u}^{+\infty} {}_{t-h-u}p_{x+h+u}^{ii}\, v^{t-h-u}\, \mathrm{d}t \right]\, \mathrm{d}u. \tag{3.134}$$

We have already noted that the integral in brackets is the expected present value of a continuous annuity which ceases on death or recovery. In the case of time-discrete annuities, the integral should be replaced by an appropriate sum. Let us denote by a_{x+h+u}^* the value of this sum. Hence, the right-hand side of (3.134) becomes a sum of terms of the following type:

$$\int_0^1 {}_{h+u}p_x^{aa}\, \mu_{x+h+u}^{ai}\, v^{h+u}\, a_{x+h+u}^*\, \mathrm{d}u \tag{3.134'}$$

for which the following approximation can be assumed:

$$v^{h+1/2}\, a_{x+h+1/2}^* \int_0^1 {}_{h+u}p_x^{aa}\, \mu_{x+h+u}^{ai}\, \mathrm{d}u. \tag{3.134''}$$

Let us consider the integral in (3.134″) with $h = 0$; it represents the expected number of transitions into disability state between ages x and $x + 1$, and it seems reasonable to assume that this integral also represents a good approximation to the probability of entering into that state between ages x and $x+1$, as the probability of two or more entries is very low. Let us denote by w_x this probability (see section 3.2.7). Now, let us return to equation (3.134) and assume for $_{h+u}p_x^{aa}$ the following approximation:

$$_{h+u}p_x^{aa} \cong {}_hp_x^{aa} \, {}_up_{x+h}^{aa}. \tag{3.135}$$

(It should be pointed out that an exact relationship is given by the Chapman–Kolmogorov equation, i.e.

$$_{h+u}p_x^{aa} = {}_hp_x^{aa} \, {}_up_{x+h}^{aa} + {}_hp_x^{ai} \, {}_up_{x+h}^{ia} \tag{3.135'}$$

from which we can observe that approximation (3.135) disregards the possibility of recovery between ages $x+h$ and $x+h+1$ for a disability commencing between those ages.)

Finally, using the approximations mentioned above, we get

$$\bar{a}_x^{ai} \cong \sum_{h=0}^{+\infty} v^{h+1/2} a_{x+h+1/2}^* \int_0^1 {}_{h+u}p_x^{aa} \, \mu_{x+h+u}^{ai} \, du \tag{3.136}$$

and then

$$\bar{a}_x^{ai} \cong \sum_{h=0}^{+\infty} {}_hp_x^{aa} w_{x+h} v^{h+1/2} a_{x+h+1/2}^*. \tag{3.137}$$

The right-hand side of (3.137) is an 'inception-annuity' formula for the single premium of a disability cover. It is a time-discrete formula, based on the probabilities w_{x+h} of becoming sick ('inception'), and on the expected present value $a_{x+h+1/2}^*$ of an annuity payable while the insured remains sick. Formula (3.137) is susceptible to adjustments and improvements, for example using select probabilities in determining the expected present value of the annuity.

Some examples of formulae used in actuarial practice follow. For the sake of brevity, we shall restrict our attention to premium calculations.

Let us start with some formulae used in the United States. Assume that the set of policy conditions is defined by $\Gamma = [0, n, s/12, \infty, \xi - x]$, where x is the age at policy entry and ξ denotes some fixed age (e.g. the retirement age). Assume that the disability annuity is paid on a monthly basis. Let $a_{x:n|\xi,s/12]}^{ai}$ denote the single premium and assume:

$$a_{x:n|\xi,s/12]}^{ai} = \sum_{h=0}^{n-1} v^h \, {}_hp_x^{aa} v^{1/2} w_{x+h} v^{s/12} \, {}_{s/12}p_{[x+h+1/2]}^{ii}$$

$$\times \ddot{a}_{[x+h+1/2]+s/12:\xi-x-h-1/2-s/12]}^{(12)i} \tag{3.138}$$

where $\ddot{a}^{(12)i}$ denotes the actuarial value of a disability annuity paid in advance on a monthly basis. Note that $x+h+1/2$ relates to the age at disablement, while $x+h+1/2+s/12$ represents the age at which the deferment ends and then the annuity commences.

It is interesting to compare (3.138) with (3.137). Some differences obviously depend on the policy conditions which we have assumed. Moreover, the probabilities of remaining disabled and hence the actuarial value of the disability annuity are assumed to be inception-select.

Note that the use of w_{x+h} (probability of becoming disabled between age $x+h$ and $x+h+1$) instead of p^{ai}_{x+h} (probability of being disabled at age $x+h+1$) can be justified by the fact that the annuity on a monthly basis is paid also to persons becoming disabled between ages $x+h$ and $x+h+1$ who are not disabled at age $x+h+1$.

In order to achieve simpler implementations, formula (3.138) can be modified in several ways. We now illustrate some modifications adopted in actuarial practice. First, the transition into the disability state is assumed to occur only if the disabled person survives to the end of the deferred period of s months; if death occurs during the deferred period, the transition out of the active state is regarded as death. Hence, the survivorship term $_{s/12}p^{ii}_{[x+h+1/2]}$ is deleted, as the survival is implicitly considered in w. A second simplification consists in using continuous annuities for disabled persons, by means of the approximation

$$\bar{a}^i_{[x+h+1/2]+s/12:\xi-x-h-1/2-s/12]} + \frac{1}{24} \qquad (3.139)$$

for the actuarial value of the disability annuity. Note that, if compared with the monthly annuity, the continuous annuity loses approximately $1/2$ month interest on each monthly payment of $1/12$. In the case of termination by death or recovery there is also a partial loss of payment (which does not occur in case of survival to age ξ in the disability state). Thus, an adjustment of the continuous annuity by $1/2$ of a monthly payment (i.e. by $1/24$) is convenient.

The third simplification is to use ordinary survival probabilities, $_hp_x$, instead of probabilities of being active, $_hp^{aa}_x$. This has the effect of increasing the actuarial value of benefits.

The fourth approximation concerns the factor w_{x+h}. The probability w_y relates to an individual aged y. Hence it must be estimated, referring the number of lives entering into the disability state between age y and $y+1$ to the number of actives (exposed to risk) aged y. In place of w_y we can use the quantity w'_y, which is defined regarding as exposed to risk the average number of actives between age y and $y+1$. The following relation holds:

$$w_y = w'_y \left(1 - \frac{1}{2}q^{aa}_y\right) \qquad (3.140)$$

which can be approximately expressed as follows:

$$w_y = w'_y \cdot {_{1/2}}p_y \qquad (3.140')$$

These simplifications lead to the following formula, known as the **Phillips approximation**:

$$a^{ai}_{x:n|\xi,s/12|} = \sum_{h=0}^{n-1} v^{h+1/2} \, {}_{h+1/2}p_x w'_{x+h} v^{s/12}$$

$$\times \left(\bar{a}^i_{[x+h+1/2]+s/12:\xi-x-h-1/2-s/12|} + \frac{1}{24} \right). \tag{3.141}$$

Formulae (3.138) and (3.141) are based on the probability of becoming disabled, w (i.e. on the probability of 'inception' of a disability spell), and the actuarial value of a disability annuity, $\bar{a}^i$. This justifies the label **inception-annuity model**, which is used to denote formulae of the type described above. Similar methods are used for the actuarial treatment of the disability risk in Germany, Austria and Switzerland, as well as in some other countries of the European continent. However, the US method differs from the European ones in the calculation of the disability annuity actuarial value. Actually, the method applied in the US market (which is also known as the **continuance table method**) is based on the probabilities for a disabled person, remaining in the disability state for a certain length of time, i.e. on a 'continuance table'. However, the method used in Germany, Austria and Switzerland (also called the **method of decrement tables**) is based on the disabled mortality rates and the recovery rates, i.e. on the rates relating to the two causes of decrement from the disability state.

Now, let us briefly describe some actuarial methods used in Europe and based on the inception-annuity model. The single premium formula used in Germany is as follows:

$$a^{ai}_{x:n|} = \sum_{h=0}^{n-1} v^h \, {}_hp^{aa}_x \left(1 - \frac{1}{2}q^{aa}_{x+h} \right) w'_{x+h} v^{1/2}$$

$$\times \frac{1}{2} \left(\ddot{a}^i_{[x+h]:n-h|} + \ddot{a}^i_{[x+h+1]:n-h-1|} \right) \tag{3.142}$$

where $\ddot{a}^i_{[x+h]:n-h|}$ denotes the inception-select actuarial value of an annuity payable (in advance) at policy anniversaries until the insured reactivates or dies and for n years at most, i.e.:

$$\ddot{a}^i_{[x+h]:n-h|} = \sum_{k=0}^{n-h-1} v^k \, {}_kp^{ii}_{[y]}. \tag{3.142'}$$

The meaning of the factor w'_{x+h} and the related presence of the factor $(1 - \frac{1}{2}q^{aa}_{x+h})$ have been already explained above. Note that a linear interpolation formula is used to determine the actuarial value of a disability annuity commencing between ages $x+h$ and $x+h+1$.

The single premium formula used in Austria is as follows:

$$a^{ai}_{x:n]} = \sum_{h=0}^{n-1} v^h \, {}_hp^{aa}_x w_{x+h} v^{1/2} \tfrac{1}{2} (\ddot{a}^i_{[x+h]:m-h]} + \ddot{a}^i_{[x+h+1]:m-h-1]}). \qquad (3.143)$$

Note that n denotes the period of cover and m the longest period of annuity payment. The factor $(1 - \tfrac{1}{2} q^{aa}_{x+h})$ is omitted since probabilities of the type w_{x+h} are used.

In Switzerland, both total and partial disability are covered. The disability benefits are scaled according to the degree of disablement. Then, a new item must be introduced in premium calculation. Actually, the quantity g_{x+h} in equation (3.144) reflects the mean degree of disability. The single premium formula is as follows:

$$a^{ai}_{x:n]} = \sum_{h=0}^{n-1} v^h \, {}_{h+1/2}p_x w'_{x+h} g_{x+h} v^{1/2} \tfrac{1}{2} (\ddot{a}^i_{[x+h]:n-h]} + \ddot{a}^i_{[x+h+1]:n-h-1]}). \quad (3.144)$$

Note the use of an ordinary mortality table.

As far as annual premiums are concerned, the following formula should be adopted (assuming that the premium payment period is equal to n):

$$P_{x:n]} = \frac{a^{ai}_{x:n]}}{\ddot{a}^{aa}_{x:n]}}. \qquad (3.145)$$

The calculation of the actuarial value $\ddot{a}^{aa}_{x:n]}$ requires the use of probabilities of being active, p^{aa}. When these probabilities are not involved in the calculation of $a^{ai}_{x:n]}$ (see formulae used in the US and Switzerland), the actuarial value $\ddot{a}^{aa}_{x:n]}$ is replaced by the actuarial value of an ordinary life annuity, i.e. $\ddot{a}_{x:n]}$. Hence, the annual premium is approximately calculated as follows:

$$P_{x:n]} = \frac{a^{ai}_{x:n]}}{\ddot{a}_{x:n]}}. \qquad (3.146)$$

Note that the same simplification, which is achieved by leaving cases of disablement out of consideration, affects the numerator as well as the denominator of (3.146) in the various implementations described earlier (see formulae (3.141) and (3.144)). It follows that the inaccuracy introduced by this simplification is not dramatic. The above-mentioned approximation problem is known in the actuarial literature as Jacob's problem (see section 3.4.5).

3.4.2 The Dutch model

In section 3.1 we briefly described the main types of disability cover sold in Holland:

1. the first-year risk cover (A-cover), with a deferment period between seven days and six months, and a maximum period of annuity payment equal to one year;
2. the after-first-year risk cover (B-cover), with a deferment period equal to one year.

We will briefly discuss the actuarial structure of the A-cover in section 3.4.4. We now illustrate the actuarial model used for pricing and reserving disability annuities of type B.

The disability actuarial model used in Holland for the B-cover constitutes an interesting implementation of the time-discrete Markov model in which the disability state is split into a given number of states according to the disability duration. In fact, a semi-Markov model is used, since duration effects are taken into account. For general aspects concerning the probabilistic structure the reader should refer to Chapter 2, section 2.3. Here we restrict our attention to some particular features of the Dutch implementation. A numerical example mainly based on the Dutch model and the relevant formulae for the calculation of actuarial values (premiums and reserves) will be presented in section 3.4.6.

The select period is assumed to be five years, after which the effect of selection is assumed to have no relevance. Hence, the disability state is split into six states. For states $i_1, i_2, \ldots, i_5$ recovery is possible (at least in principle), whilst i_6 is assumed to represent permanent disability. The Dutch life insurers assume the following simplifying hypothesis about the transition probabilities:

$$p_y^{i_s i_1} = 0; \quad s = 1, 2, \ldots, 6; \tag{3.147}$$

thus, it is assumed that it is not possible to reactivate and become disabled again in the same year.

The above-mentioned assumptions lead to the one-year transition matrix of Table 3.1.

Moreover, it is assumed that both the disabled mortality and the active mortality are equal to general population mortality:

$$p_y^{i_s d} = p_y^{ad} = q_y; \quad s = 1, 2, \ldots, 6; \tag{3.148}$$

Note, however, that the overall probability of death, q_y, is not well-defined within the context of the Markov model, since it is independent of the state at age y.

The probabilities of disablement, i.e. $p_y^{ai_1}$, are assumed to be an exponential function of the attained age, y:

$$p_y^{ai_1} = 0.00223 \times 1.0468^y. \tag{3.149}$$

The probabilities of recovery are assumed to be decreasing linear functions of the attained age for each s; more precisely, the following functions

Table 3.1 One-year transition probabilities

	a	i_1	i_2	i_3	i_4	i_5	i_6	d
a	p_y^{aa}	$p_y^{ai_1}$	0	0	0	0	0	p_y^{ad}
i_1	$p_y^{i_1a}$	0	$p_y^{i_1i_2}$	0	0	0	0	$p_y^{i_1d}$
i_2	$p_y^{i_2a}$	0	0	$p_y^{i_2i_3}$	0	0	0	$p_y^{i_2d}$
i_3	$p_y^{i_3a}$	0	0	0	$p_y^{i_3i_4}$	0	0	$p_y^{i_3d}$
i_4	$p_y^{i_4a}$	0	0	0	0	$p_y^{i_4i_5}$	0	$p_y^{i_4d}$
i_5	$p_y^{i_5a}$	0	0	0	0	0	$p_y^{i_5i_6}$	$p_y^{i_5d}$
i_6	0	0	0	0	0	0	$p_y^{i_6i_6}$	$p_y^{i_6d}$
d	0	0	0	0	0	0	0	1

are used:

$$p_y^{i_sa} = \max\{\alpha_s - \beta_s y; 0\}; \quad s = 1, 2, 3, 4, 5. \tag{3.150}$$

The parameters a_s, β_s are given by Table 3.2.

Note that from (3.150) we obtain probabilities of recovery equal to zero for high ages and/or high disability durations.

For premium and reserve calculations, formulae consistent with the Dutch model will be presented in section 3.4.6.

3.4.3 The Manchester Unity model

The 'Manchester Unity' model (or 'Friendly Society' model) is the method that has been traditionally used in the UK, until the publication of CMIR 12 in 1991, for calculating actuarial values in disability insurance. It is interesting to reinterpret (and criticize) the relevant formulae in the framework of the Markovian multiple state model. For brevity, we shall restrict our attention to the probabilistic structure and single premium calculation.

Let us now consider the probability f_{x+t} that an individual is disabled at age $x + t$ given that he is alive at age $x + t$ (disregarding the state at age x,

Table 3.2 Parameters for probabilities of recovery

s	α_s	β_s
1	1.24111	0.02219
2	0.66499	0.01153
3	0.27394	0.00532
4	0.23547	0.00470
5	0.14166	0.00319

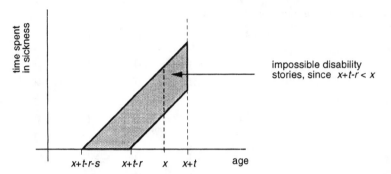

Figure 3.12 Impossible disability stories.

that is at policy issue):

$$f_{x+t} = \Pr\{S(x+t) = i \mid (S(x+t) = a) \vee (S(x+t) = i)\}. \qquad (3.151)$$

Note that the above-mentioned probability is not well-defined within the Markov model, since no previous information is taken into account (in particular, the state at age x being disregarded). A reinterpretation of f_{x+t} will be proposed later.

Let us assume the following approximation (similar to equation (3.88)) for the single premium:

$$\bar{a}^{ai}_{x:n|} \cong \int_0^n {}_tp_x f_{x+t} v^t \, dt = \int_0^n \frac{l_{x+t}}{l_x} f_{x+t} v^t \, dt \qquad (3.152)$$

from which it appears that this technique is also based on the probability of being disabled.

It must be pointed out that while in equation (3.88) the probability $j_{(x)+t}$ is a function of the two variables x and t, the probability f_{x+t} used in (3.152) depends only on the attained age $x + t$. It follows that for an insured aged $x + t$ and disabled at that age, any previous age is considered as a possible age of inception of the current disability spell, whilst only disabilities commencing after the entry age x can really lead to the current disability spell. Figure 3.12 illustrates a set of impossible disability stories (those commencing between ages $x + t - r - s$ and $x + t - r$, with $r > t$, $s > 0$) for an individual aged $x + t$ who purchased the insurance cover at age x. A significant overestimate of probabilities f_{x+t} might occur when the estimation of these probabilities is based on claims from a 'mature' portfolio.

For the single premium $\bar{a}^{ai}_{x:n|}$, we can adopt the following approximation:

$$\int_0^n \frac{l_{x+t}}{l_x} f_{x+t} v^t \, dt = \sum_{h=1}^n \frac{1}{l_x} \int_0^1 f_{x+h-1+\tau} \, l_{x+h-1+\tau} \, v^{h-1+\tau} \, d\tau$$

$$\cong \sum_{h=1}^n \frac{l_{x+h-1/2}}{l_x} v^{h-1/2} \frac{\int_0^1 f_{x+h-1+\tau} \, l_{x+h-1+\tau} \, d\tau}{\int_0^1 l_{x+h-1+\tau} \, d\tau}. \qquad (3.153)$$

Defining the function

$$\theta_y = \frac{\int_0^1 f_{y+\tau}\, l_{y+\tau}\, d\tau}{\int_0^1 l_{y+\tau}\, d\tau} \tag{3.154}$$

which represents the ratio of the expected time spent in sickness between ages y and $y+1$ to the expected time lived between ages y and $y+1$ (the **central sickness rate**), we obtain the more usual formula for the single premium:

$$\bar{a}^{ai}_{x:n]} \cong \sum_{h=1}^{n} \frac{l_{x+h-1/2}}{l_x}\, v^{h-1/2}\theta_{x+h-1}. \tag{3.155}$$

Formula (3.155) can be interpreted (as an approximate formula) in the light of the Markov modelling. Actually, the formula can be compared with equation (1.120); note that a simple survival probability is now used and that the 'central' sickness rate θ_{x+h-1} refers to an insured alive at age $x+h-\frac{1}{2}$.

Formulae useful in practice can be implemented through the probabilities $f_y^{r/s}$ that an insured aged y is disabled, with duration of disability between r and $r+s$. According to the previous graphical convention, the shaded region in Fig. 3.13 represents the set of disability stories which imply disability at age y with duration between r and $r+s$.

The probability $f_y^{r/s}$ can approximately be expressed in terms of probabilities ${}_tp_{x_0}^{aa}$, ${}_tp_{x_0}^{ai}$ and ${}_tp_{x_0}^{ai}(\tau)$ (see equations (3.7) and (3.10)), where x_0 denotes an age at which the individual was active. To this purpose, let us consider the probability that an individual, active at age x_0, is alive at age y,

$$\Pr\{(S(y)=a)\vee(S(y)=i)\,|\,S(x_0)=a\} = {}_{y-x_0}p_{x_0}^{aa} + {}_{y-x_0}p_{x_0}^{ai} \tag{3.156}$$

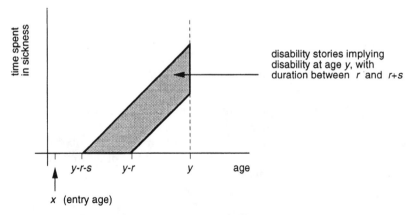

Figure 3.13 Disability stories with duration between r and $r+s$.

and the probability that he is disabled at age y with disability duration between r and $r + s$

$$\Pr\{S(x_0 + u) = i \text{ for all } u \in [y - x_0 - k, y - x_0]$$
$$\text{with } r \leq k \leq r + s \mid S(x_0 = a)\}$$
$$= {}_{y-x_0}p_{x_0}^{ai}(r) - {}_{y-x_0}p_{x_0}^{ai}(r + s). \tag{3.157}$$

Let $f_{y,x_0}^{r/s}$ denote the probability that the insured, alive at age y, is disabled at that age with disability duration between r and $r + s$, given that he was active at some age x_0, $x_0 < y - (r + s)$; it follows:

$$f_{y,x_0}^{r/s} = \frac{{}_{y-x_0}p_{x_0}^{ai}(r) - {}_{y-x_0}p_{x_0}^{ai}(r + s)}{{}_{y-x_0}p_{x_0}^{aa} + {}_{y-x_0}p_{x_0}^{ai}}. \tag{3.158}$$

The probability $f_y^{r/s}$ can be considered as an approximation to $f_{y,x_0}^{r/s}$. In particular:

$$f_y = f_y^{0/\infty} \cong \frac{{}_{y-x_0}p_{x_0}^{ai}}{{}_{y-x_0}p_{x_0}^{aa} + {}_{y-x_0}p_{x_0}^{ai}}. \tag{3.159}$$

3.4.4 A model for short-term sickness benefits

By 'short-term sickness benefits' we mean disability benefits provided in the case of acute cases of sickness usually lasting no longer than some months. Benefits are paid after a short deferred period (say some weeks) and cease after a given period, usually six or 12 months even though the disability lasts longer. Premiums are often calculated on a one-year basis, using methods coming from non-life actuarial mathematics.

First, let us consider the one-year premium in the context of the time-continuous Markov modelling. We then have:

$$\bar{a}_{x:1|}^{ai} = \int_0^1 v^u \, {}_u p_x^{ai} \, du. \tag{3.160}$$

Whence

$$\bar{a}_{x:1|}^{ai} \cong v^{1/2} \int_0^1 {}_u p_x^{ai} \, du = v^{1/2} \bar{e}_{x:1|}^{ai} \tag{3.161}$$

where $\bar{e}_{x:1|}^{ai}$ denotes the expected time spent in the disability state between ages x and $x + 1$ by an insured who is active at age x.

In the case of deferment, policy conditions are described by $\Gamma = [0, 1, f, \infty, 1]$ (see section 3.2.5). Hence, the one-year premium should be correctly calculated as follows:

$$\bar{a}_{x,\Gamma}^{ai} = \int_f^1 v^u \, {}_u p_x^{ai}(f) \, du. \tag{3.162}$$

However, in actuarial practice it is usual to resort to approximation formulae. In particular, the effect of deferment can be dealt with by using a reduction parameter. Further 'risk factors' can also be introduced by using appropriate formulae. Multiplicative models are often used to this purpose. When these models are used, an overall basic premium is calculated, say P. Then, the effects of various risk elements are expressed by using appropriate multiplicative factors.

The disability A-cover sold by Dutch insurers (see section 3.1.3) represents an example of short-term sickness benefits. The calculation of a (one-year) premium is based on a multiplicative model:

$$P(i,j,k) = P\rho(i)\beta(j)\gamma(k) \tag{3.163}$$

where the parameters relate to the three risk factors considered in the Dutch implementation, i.e. deferred period (five possibilities), age (nine groups), profession (four classes).

3.4.5 Jacob's problem

Let us consider an inception/annuity formula for the calculation of a single premium, for example (see equation (3.137)):

$$\bar{a}_{x:n]}^{ai} \cong \sum_{h=0}^{n-1} {}_hp_x^{aa} w_{x+h} v^{h+1/2} a_{x+h+1/2}^*. \tag{3.164}$$

Assuming that the premium payment period is equal to n, formula (3.145) should be used to calculate the annual premium $P_{x:n]}$. Hence, we obtain:

$$P_{x:n]} = \frac{a_{x:n]}^{ai}}{\ddot{a}_{x:n]}^{aa}} \cong \frac{\sum_{h=0}^{n-1} {}_hp_x^{aa} v^h w_{x+h} v^{1/2} a_{x+h+1/2}^*}{\sum_{h=0}^{n-1} {}_hp_x^{aa} v^h} \tag{3.165}$$

Note that the denominator is the actuarial value of an annuity paid by the insured while active, as premiums are assumed to be waived during disability spells. In practice (see section 3.4.1) $\ddot{a}_{x:n]}^{aa}$ is usually replaced by $\ddot{a}_{x:n]}$. Moreover, the same simplification often affects the numerator as well (see, for example, equations (3.141) and (3.144)). Then, we obtain a different premium:

$$P'_{x:n]} = \frac{\sum_{h=0}^{n-1} {}_hp_x v^h w_{x+h} v^{1/2} a_{x+h+1/2}^*}{\sum_{h=0}^{n-1} {}_hp_x v^h}. \tag{3.166}$$

Formula (3.166) constitutes a reasonable approximation, and this can be illustrated by numerical computations. However, formulae (3.165) and (3.166) can be compared also in a formal context. For the sake of brevity, we shall restrict our discussion to a simple outline of this problem.

From equations (3.165) and (3.166) we see that $P_{x:n]}$ and $P'_{x:n]}$ can be interpreted as weighted averages of the 'natural' premiums given by $w_{x+h}v^{1/2}a^*_{x+h+1/2}$, for $h = 0, 1, \ldots, n-1$. The second formula differs from the first one in one respect only, i.e. the substitution of $_hp_x$ for $_hp_x^{aa}$. Further weighting systems can be considered, at least in principle. For example:

$$P''_{x:n]} = \frac{\sum_{h=0}^{n-1} v^h w_{x+h} v^{1/2} a^*_{x+h+1/2}}{\sum_{h=0}^{n-1} v^h}. \tag{3.167}$$

This formula represents a much more substantial simplification as it completely neglects termination probabilities. Nevertheless, formula (3.167) also provides a reasonable approximation to the correct premium $P_{x:n]}$.

The effects of the different weighting systems can be analysed by examining the actuarial content of the related formulae. M. Jacob studied this problem in 1934 (see Jacob, 1934). In particular, he proved that the differences between the premiums are merely equal to the premiums for rider benefits having as sum assured the mathematical reserves relating to both the disability insurance and the rider benefit itself. For the most common ages and durations these reserves are relatively small, whence the conclusion that the differences are negligible if compared with the required premiums. These results also indicate the small influence exerted by the mortality of active lives at the most common ages on the level of annual premiums.

3.4.6 A numerical example

We now present a numerical example, in order to illustrate premiums and reserves as calculated in a time-discrete Markov context. The disability duration effect is taken into account by splitting the disability state, as in the Dutch model. Hence, an inception-select scheme is actually adopted as far as recovery is concerned. In particular, probabilities of disablement and probabilities of recovery presented in section 3.4.2 are used (see in particular equations (3.149) and (3.150) and Table 3.2). Probabilities of death for active lives, p_y^{ad}, are derived from the intensity of mortality adopted by the Danish model (see section 3.4.1). Probabilities of death for disabled insureds, $p_y^{i,d}$, are assumed to be independent of s and equal to $(1 + \eta)p_y^{ad}$ (thus, a non-inception-select scheme has been used for the mortality of disabled lives). The technical rate of interest $i = 0.04$ has been assumed.

In order to perform calculations according to the probabilistic structure of the Dutch model, some definitions of actuarial values must be generalized.

First, let $\ddot{a}^{aa}_{x:\overline{n}|}$ denote as usual the actuarial value of an 'active' annuity:

$$\ddot{a}^{aa}_{x:\overline{n}|} = \sum_{k=0}^{n-1} {}_k p^{aa}_x v^k. \qquad (3.168)$$

The disability benefit (1 monetary unit per annum) is paid whatever disability state i_s, $s = 1, 2, \ldots, 6$, is occupied by the insured. Hence, the actuarial value of the disability benefits is defined, for an active individual, as follows:

$$a^{ai}_{x:\overline{n}|} = \sum_{k=1}^{n} \sum_{s=1}^{6} {}_k p^{ai_s}_x v^k. \qquad (3.169)$$

Note that the disability annuity is assumed to be payable up to the end of the policy term n. This seems to be more realistic than assuming the last payment at time $n - 1$, as in insurance practice monthly benefits are usually arranged.

Conversely, the actuarial value of the disability benefits is defined, for a disabled individual aged $x + t$ who occupies the state i_s ($s = 1, \ldots, 6$), as follows:

$$\ddot{a}^{i_s i}_{x+t:\overline{n-t+1}|} = \sum_{k=0}^{n-t} \sum_{h=1}^{6} {}_k p^{i_s i_h}_{x+t} v^k, \quad t = 1, \ldots, n \qquad (3.170)$$

where ${}_0 p^{i_s i_h}_{x+t} = 1$ if $h = s$ and 0 otherwise.

Since the disability annuity is assumed to be payable until the policy term n, in equation (3.170) the actuarial value of the disability benefits refers to $n - t + 1$ payments.

The annual premium, $P_{x:\overline{n}|}$, payable for n years while the insured is active, is given by:

$$P_{x:\overline{n}|} = \frac{a^{ai}_{x:\overline{n}|}}{\ddot{a}^{aa}_{x:\overline{n}|}}. \qquad (3.171)$$

The calculation of annual premiums payable for m years, $m < n$, is straightforward.

The active reserve (when premiums are payable for n years) is given by:

$$V^a_{x+t,\,m-t} = a^{ai}_{x+t:\overline{n-t}|} - P_{x:\overline{n}|} \ddot{a}^{aa}_{x+t:\overline{n-t}|}, \quad t = 0, \ldots, n, \qquad (3.172)$$

whereas the disabled reserve, relating to state i_s ($s = 1, 2, \ldots, 6$), is calculated (in practice) as follows:

$$V^{i_s}_{x+t,\,m-t} = \ddot{a}^{i_s i}_{x+t:\overline{n-t+1}|}, \quad t = 1, \ldots, n. \qquad (3.173)$$

Note that equation (3.173) neglects the actuarial value of premiums following the possible recovery, whilst the possibility of recovery is implicitly taken into account by the probabilities ${}_k p^{i_s i_h}_{x+t}$ upon which the actuarial value $\ddot{a}^{i_s i}_{x+t:\overline{n-t+1}|}$ is based. In the following numerical examples we will disregard the possibility of recovery when calculating the disabled reserve. This assumption is consistent with the simplification commonly adopted in actuarial practice, already mentioned (see, for example, formula (3.81)).

The calculation of the actuarial values defined by equations (3.168), (3.169) and (3.170) and then the calculation of premiums and reserves according to formulae (3.171), (3.172) and (3.173) obviously imply the use of the underlying probabilities, which can be derived from the one-year transition matrix (see Table 3.1). Otherwise, actuarial values can be calculated by using recursion formulae. Recursion formulae, which can be easily interpreted by direct reasoning, are as follows:

$$\ddot{a}^{aa}_{x+t:n-t]} = 1 + p^{aa}_{x+t} v \ddot{a}^{aa}_{x+t+1:n-t-1]} + p^{ai_1}_{x+t} v \ddot{a}^{i_1a}_{x+t+1:n-t-1]} \tag{3.174}$$

$$\ddot{a}^{i_sa}_{x+t:n-t]} = p^{i_sa}_{x+t} v \ddot{a}^{aa}_{x+t+1:n-t-1]} + p^{i_si_{s+1}}_{x+t} v \ddot{a}^{i_{s+1}a}_{x+t+1:n-t-1]}$$

$$s = 1, 2, \ldots, 6 \tag{3.175}$$

(with $\ddot{a}^{i_6a} = 0$ for all ages and residual durations)

$$a^{ai}_{x+t:n-t]} = p^{aa}_{x+t} v a^{ai}_{x+t+1:n-t-1]} + p^{ai_1}_{x+t} v \ddot{a}^{i_1i}_{x+t+1:n-t]} \tag{3.176}$$

$$\ddot{a}^{i_si}_{x+t:n-t+1]} = 1 + p^{i_sa}_{x+t} v a^{ai}_{x+t+1:n-t-1]} + p^{i_si_{s+1}}_{x+t} v \ddot{a}^{i_{s+1}i}_{x+t+1:n-t]}$$

$$s = 1, 2, \ldots, 5 \tag{3.177}$$

$$\ddot{a}^{i_6i}_{x+t:n-t+1]} = 1 + p^{i_6i_6}_{x+t} v a^{i_6i}_{x+t+1:n-t]} \tag{3.178}$$

(with $\ddot{a}^{i_6i}_{x+n:1]} = 1$).

It should be stressed that formulae (3.174) to (3.178) are only based on the one-year transition probabilities.

The consideration of natural premiums can help in interpreting some results concerning the behaviour of the active reserve. To this purpose, let us define the following probability:

$$_r p^{i_1i}_{x+t} = \prod_{h=1}^{r} p^{i_hi_{h+1}}_{x+t+h-1} \tag{3.179}$$

with $i_h = 6$ if $h > 6$. Then, let us define

$$\ddot{a}^{\overline{i_1i}}_{x+t:n-t+1]} = 1 + v_1 \, p^{\overline{i_1i}}_{x+t} + v^2 \, _2p^{\overline{i_1i}}_{x+t} + \cdots \tag{3.180}$$

the sum consisting of $n - t + 1$ terms. The natural premium is defined as follows:

$$P^{(N)}_t = v p^{ai_1}_{x+t-1} \ddot{a}^{\overline{i_1i}}_{x+t:n-t+1]}. \tag{3.181}$$

The natural premium $P^{(N)}_t$ represents the expected cost at time $t - 1$, i.e. at the beginning of the tth year, of the disability spell which might commence during the year itself.

Table 3.3 Probabilities of disablement $p_y^{ai_1}$

$y = 30$	$y = 45$	$y = 55$	$y = 60$
0.00879	0.01747	0.02759	0.03468

It is possible to prove that the actuarial value $a^{ai}_{x:n\rceil}$ can be expressed as follows:

$$a^{ai}_{x:n\rceil} = \sum_{k=1}^{n} {}_{k-1}p^{aa}_{x} p^{ai_1}_{x+k-1} v^{k} \ddot{a}^{\overline{i_1 i}}_{x+k\,:\,n-k+1\rceil} \tag{3.182}$$

and hence, in terms of natural premiums:

$$a^{ai}_{x:n\rceil} = \sum_{k=1}^{n} {}_{k-1}p^{aa}_{x} v^{k-1} P^{(N)}_{k}. \tag{3.183}$$

Formula (3.182) follows from

$$_{k}p^{ai_s}_{x} = \sum_{h=1}^{k} {}_{h-1}p^{aa}_{x} p^{ai_1}_{x+h-1} \cdot {}_{k-h}p^{\overline{i_1 i}}_{x+h} \tag{3.184}$$

which in turn can be derived from the Chapman–Kolmogorov equations.

Using (3.168), (3.171) and (3.183) we see that the annual premium $P_{x:n\rceil}$ payable for n years can be expressed as a weighted arithmetic mean of the natural premiums. This result (well-known in life insurance mathematics) can help in interpreting the behaviour of the active reserve.

Now, let us turn to numerical examples. Table 3.3 illustrates probabilities of disablement at various ages, calculated according to formula (3.149). Probabilities of recovery, calculated according to (3.150) with the parameters specified in Table 3.2, are shown in Table 3.4.

The adopted probabilities obviously affect actuarial values such as premiums and reserves. In particular, the results we now present notably differ from the corresponding results relating to the Danish model. This is due to the probability values which in turn depend on the policy conditions. In general, a dramatic effect can be attributed to the percentage of disability which entitles the insured to benefits.

Table 3.4 Probabilities of recovery $p_y^{i,a}$

s	$y = 30$	$y = 45$	$y = 55$	$y = 60$
1	0.57541	0.24256	0.02066	0.00000
2	0.31909	0.14614	0.03084	0.00000
3	0.11434	0.03454	0.00000	0.00000
4	0.09447	0.02397	0.00000	0.00000
5	0.04596	0.00000	0.00000	0.00000
6	0.00000	0.00000	0.00000	0.00000

Table 3.5 Single and annual premiums, $\eta = 0.30$, $m = n$

	$x = 30$	$x = 40$	$x = 50$
$n = 10$	0.20848	0.44950	0.89054
	0.02531	0.05612	0.11811
$n = 15$	0.41786	0.92343	1.73095
	0.03752	0.08696	0.18010
$n = 20$	0.71079	1.54000	2.61699
	0.05310	0.12351	0.24075

Table 3.6 Single and annual premiums, $\eta = 0$, $m = n$

	$x = 30$	$x = 40$	$x = 50$
$n = 10$	0.20860	0.45017	0.89478
	0.02533	0.05620	0.11868
$n = 15$	0.41831	0.92620	1.74866
	0.03756	0.08722	0.18194
$n = 20$	0.71211	1.54865	2.66750
	0.05320	0.12421	0.24540

Single and annual premiums corresponding to $\eta = 0.30$ and $\eta = 0$ are shown in Tables 3.5 and 3.6 respectively. Annual premiums are assumed to be payable for $m = n$ years.

Figure 3.14 illustrates the behaviour of the single premium as a function of the extra level of mortality η. In the following calculation (some exceptions apart) we have assumed $\eta = 0.30$.

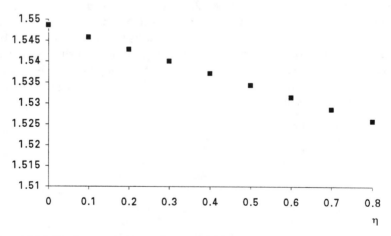

Figure 3.14 Single premium as a function of the extra level of mortality η; $x = 40$, $n = 20$.

Table 3.7 Natural premiums, $\eta = 0.30$, $n = 20$

	$x = 30$	$x = 40$	$x = 50$
0	0.03866	0.10268	0.23325
1	0.04173	0.10912	0.24346
2	0.04487	0.11553	0.25327
3	0.04804	0.12183	0.26256
4	0.05119	0.12792	0.27112
5	0.05428	0.13367	0.27837
6	0.05724	0.13894	0.27838
7	0.06000	0.14319	0.27536
8	0.06246	0.14654	0.27094
9	0.06452	0.14855	0.26495
10	0.06600	0.14898	0.25720
11	0.06678	0.14784	0.24745
12	0.06671	0.14479	0.23543
13	0.06560	0.13948	0.22083
14	0.06320	0.13148	0.20328
15	0.05926	0.12014	0.18230
16	0.05345	0.10333	0.15732
17	0.04544	0.08303	0.12765
18	0.03487	0.05935	0.09238
19	0.02016	0.03186	0.05033

The behaviour of natural premiums is illustrated by Table 3.7. Natural premiums initially increase as duration since policy issue increases, then decrease. Formula (3.181) shows that the natural premium is the product of a factor, $p^{ai_1}_{x+t-1}$, increasing as the attained age increases, and a factor, $\ddot{a}^{i_1 i}_{x+t:n-t+1]}$, decreasing as the attained age increases, because the possible residual duration of the disability annuity decreases. Hence, the behaviour of natural premiums can be interpreted as the combined result of the two factors above mentioned.

The behaviour of natural premiums implies negative values of the active reserve, as shown in Table 3.8. Hence, it is necessary to shorten the premium payment period, m. Table 3.9 illustrates annual premiums corresponding to $m < n$. Table 3.10 exhibits the behaviour of the active reserve for $n = 20$ and $m = 15$.

Table 3.11 illustrates disabled reserves corresponding to various ages at disability inception. The behaviour of the disabled reserve can be easily interpreted within the context of an inception-select scheme. Table 3.12 illustrates the effect of the extra level of mortality on the disabled reserve.

Tables 3.13, 3.14 and 3.15 concern a disability cover according to which the disability annuity is payable up to a 'limit' age ξ (for example the

Table 3.8 Active reserve, $\eta = 0.30$, $m = n = 20$

t	$x = 30$	$x = 40$	$x = 50$
0	0.00000	0.00000	0.00000
1	0.01497	0.02172	0.00800
2	0.02742	0.03785	0.00571
3	0.03717	0.04815	−0.00718
4	0.04408	0.05245	−0.03091
5	0.04806	0.05068	−0.06564
6	0.04905	0.04285	−0.11103
7	0.04703	0.02910	−0.16023
8	0.04209	0.01013	−0.21048
9	0.03439	−0.01349	−0.26063
10	0.02419	−0.04071	−0.30927
11	0.01196	−0.07020	−0.35470
12	−0.00167	−0.10055	−0.39478
13	−0.01594	−0.12994	−0.42690
14	−0.02982	−0.15604	−0.44780
15	−0.04199	−0.17588	−0.45335
16	−0.05078	−0.18553	−0.43828
17	−0.05404	−0.17823	−0.39582
18	−0.04919	−0.14880	−0.31711
19	−0.03294	−0.09166	−0.19042
20	0.00000	0.00000	0.00000

retirement age), with $\xi > x + n$. According to the notation used in sections 3.2 and 3.3, we have $\xi - x = r$. Single and annual premiums are shown in Table 3.13, whereas Tables 3.14 and 3.15 illustrate the behaviour of natural premiums and of the active reserve respectively. Note that, as a consequence of the longer disability annuity payment period, natural premiums are increasing throughout the policy duration for $x = 30$ and decreasing only close to maturity for $x = 40$. The corresponding effect on the active reserve is illustrated by Table 3.15.

Finally, Figs 3.15 and 3.16 allow for comparisons among natural premiums and active reserves respectively, corresponding to different policy conditions.

Table 3.9 Annual premiums, $\eta = 0.30$, $m < n$

	$x = 30$	$x = 40$	$x = 50$
$n = 10$, $m = 7$	0.03399	0.07455	0.15336
$n = 15$, $m = 10$	0.05074	0.11529	0.22958
$n = 20$, $m = 15$	0.06382	0.14503	0.27228

Table 3.10   Active reserve, $\eta = 0.30$, $n = 20$, $m = 15$

t	$x = 30$	$x = 40$	$x = 50$
0	0.00000	0.00000	0.00000
1	0.02605	0.04407	0.04134
2	0.05006	0.08367	0.07463
3	0.07188	0.11867	0.09989
4	0.09141	0.14903	0.11724
5	0.10859	0.17481	0.12698
6	0.12339	0.19619	0.13001
7	0.13587	0.21350	0.13355
8	0.14613	0.22764	0.14095
9	0.15442	0.23945	0.15409
10	0.16106	0.25025	0.17527
11	0.16660	0.26171	0.20729
12	0.17176	0.27564	0.25357
13	0.17742	0.29433	0.31835
14	0.18483	0.32079	0.40689
15	0.19571	0.35872	0.52580
16	0.14425	0.25661	0.38125
17	0.09616	0.16520	0.24978
18	0.05382	0.08877	0.13698
19	0.02016	0.03186	0.05033
20	0.00000	0.00000	0.00000

Table 3.11   Disabled reserve, $\eta = 0.30$, $x = 40$, $n = 20$, $y =$ age at disability inception

Attained age	$y = 45$	$y = 50$	$y = 55$
46	7.95994		
47	9.31487		
48	9.89328		
49	9.45481		
50	8.87030		
51	8.21777	7.05801	
52	7.54022	7.11014	
53	6.83605	6.83605	
54	6.10400	6.10400	
55	5.34216	5.34216	
56	4.54851	4.54851	4.52814
57	3.72067	3.72067	3.69931
58	2.85571	2.85571	2.85571
59	1.95021	1.95021	1.95021
60	1.00000	1.00000	1.00000

Table 3.12 Disabled reserve, $x = 40$, $n = 20$; age at disability inception $y = 45$

Attained age	$\eta = 0$	$\eta = 0.8$
46	8.00912	7.87908
47	9.37358	9.21833
48	9.95565	9.79066
49	9.51396	9.35743
50	8.92500	8.78022
51	8.26727	8.13620
52	7.58406	7.46791
53	6.87390	6.77357
54	6.13554	6.05189
55	5.36724	5.30068
56	4.56716	4.51763
57	3.73317	3.69995
58	2.86270	2.84408
59	1.95282	1.94585
60	1.00000	1.00000

3.5 PERMANENT DISABILITY LUMP SUMS

In section 3.1 we have described some insurance policies providing annuity benefits in case of (permanent or non-necessarily permanent) disability. The relevant actuarial structure has been dealt with in sections 3.2 to 3.4. In this section we deal with covers providing a lump sum in case of permanent (and total) disability.

It must be pointed out the presence of moral hazard in this type of policy design, which involves the payment of a lump sum to an individual who may subsequently recover (fully or partially), so that the benefits are irrecoverable in this event.

We first consider an actuarial model for an n-year stand-alone cover providing a lump sum in case of permanent (and total) disability. Later we illustrate an actuarial model for a permanent disability lump sum as a rider benefit for a life assurance contract.

Table 3.13 Single and annual premiums, $\eta = 0.30$, $m = n$, $\xi = 65$

	$x = 30$	$x = 40$
$n = 10$	0.60446	1.20458
	0.07334	0.15039
$n = 15$	0.98599	1.74215
	0.08853	0.16407
$n = 20$	1.37931	2.10085
	0.10304	0.16850

Table 3.14 Natural premiums, $\eta = 0.30$, $n = 20$, $\xi = 65$

	$x = 30$	$x = 40$
0	0.04905	0.11513
1	0.05399	0.12348
2	0.05929	0.13209
3	0.06495	0.14091
4	0.07100	0.14987
5	0.07743	0.15892
6	0.08424	0.16795
7	0.09144	0.17635
8	0.09902	0.18441
9	0.10697	0.19166
10	0.11513	0.19792
11	0.12348	0.20337
12	0.13209	0.20779
13	0.14091	0.21096
14	0.14987	0.21257
15	0.15892	0.21203
16	0.16795	0.20500
17	0.17635	0.19472
18	0.18441	0.18223
19	0.19166	0.16723

Table 3.15 Active reserve, $\eta = 0.30$, $m = n = 20$, $\xi = 65$

t	$x = 30$	$x = 40$
0	0.00000	0.00000
1	0.05589	0.05555
2	0.10910	0.10521
3	0.15919	0.14852
4	0.20568	0.18503
5	0.24808	0.21433
6	0.28586	0.23603
7	0.31846	0.24979
8	0.34530	0.25587
9	0.36577	0.25422
10	0.37924	0.24524
11	0.38519	0.22950
12	0.38308	0.20739
13	0.37225	0.17951
14	0.35205	0.14671
15	0.32182	0.11012
16	0.28087	0.07162
17	0.22853	0.03785
18	0.16466	0.01256
19	0.08862	−0.00127
20	0.00000	0.00000

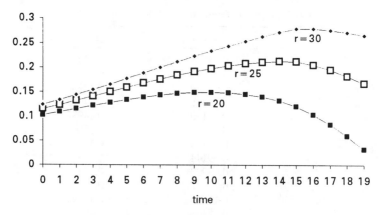

Figure 3.15 Natural premiums $x = 40$, $n = 20$.

3.5.1 Stand-alone cover

Let us start with the actuarial structure of the stand-alone cover. We assume that a qualification period of τ years is requested in order to ascertain the permanent character of the disability (usually $0.5 \leq \tau \leq 2$). In other words, the sum assured will be paid after a time τ from disability inception. The length of the qualification period should be chosen in such a way that recovery is practically impossible after that period; formally, this can be expressed in terms of the inception-select transition intensity:

$$\mu_{y,u}^{ia} = 0 \quad \text{for } u \geq \tau. \tag{3.185}$$

Then the net single premium, $\bar{A}_{x,n(\tau)}^{(PD)}$, for the permanent disability (PD) benefit is given by

$$\bar{A}_{x,n(\tau)}^{(PD)} = \int_0^n {}_t p_x^{aa} \mu_{x+t}^{ai} \cdot {}_\tau p_{x+t,0}^{ii} v^{t+\tau} \, dt. \tag{3.186}$$

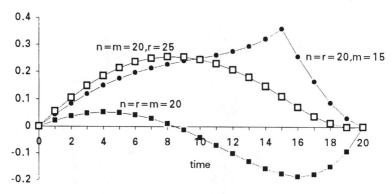

Figure 3.16 Active reserves $x = 40$.

Note that in general the integral at the right-hand side of equation (3.186) represents the actuarial value of a random benefit consisting of one monetary unit at each transition $a \to i$, if the insured remains disabled for at least a time τ after disablement. Then, relation (3.186) allows for repeated payments of the sum assured, whilst the PD cover admits one payment at most. Nevertheless, relation (3.186) holds thanks to condition (3.185), which makes it impossible for there to be a recovery after τ years from disablement.

If we choose $\tau = 0$, then the single premium is given by

$$\bar{A}_{x,n(0)}^{(PD)} = \int_0^n {}_tp_x^{\overline{aa}}\mu_{x+t}^{ai}v^t \, dt. \tag{3.187}$$

Note that in this case ${}_tp_x^{\overline{aa}}$ must be replaced by ${}_tp_x^{\overline{aa}}$: recovery is possible, at least in principle, so that the payment of the benefit should be restricted to the first transition into the disability state.

Let us return to equation (3.186). If we define

$$\sigma_\tau(x+t) = \mu_{x+t}^{ai} \cdot {}_\tau p_{x+t,0}^{\overline{ii}} \tag{3.188}$$

we can rewrite (3.186) as follows:

$$\bar{A}_{x,n(\tau)}^{(PD)} = \int_0^n {}_tp_x^{\overline{aa}}\sigma_\tau(x+t)v^{t+\tau} \, dt; \tag{3.186'}$$

note that $\sigma_\tau(x+t)\,dt$ is the probability that a healthy life aged $x+t$ becomes (permanently) disabled between ages $x+t$ and $x+t+dt$ and is still alive (and disabled) at age $x+t+\tau$. Finally, if we replace the (correct) probability ${}_tp_x^{\overline{aa}}$ by a simple survival probability ${}_tp_x$, we get the following approximation, often used in actuarial practice:

$$\bar{A}_{x,n(\tau)}^{(PD)} \cong \int_0^n {}_tp_x\sigma_\tau(x+t)v^{t+\tau} \, dt. \tag{3.189}$$

3.5.2 Rider benefit

Let us now describe an actuarial model for the permanent disability lump sum as a rider benefit for an endowment assurance. In this contract, the sum assured shall be paid upon the first event to occur among disablement, death and maturity.

Hence, the rider benefit is an 'acceleration' benefit. Thanks to this aspect of the disability cover, the qualification period can be dropped out. Then, the single premium of the contract which includes the endowment assurance as well as the disability benefit, $\bar{A}_{x,n}^{(E+PD)}$, is as follows:

$$\bar{A}_{x,n}^{(E+PD)} = \int_0^n {}_tp_x^{\overline{aa}}(\mu_{x+t}^{ai} + \mu_{x+t}^{ad})v^t \, dt + {}_np_x^{\overline{aa}}v^n. \tag{3.190}$$

Note that the last term relates to the pure endowment item in the main insurance.

The calculation of the actuarial value of the benefits in this insurance cover involves a multiple-state model of the type described in Example 5 of section 1.3, i.e. a model with two causes of decrements. Actually, the relevant transitions are $a \to i$ and $a \to d$ only, as clearly appears in formula (3.190), and no second-order decrement is involved (i.e. $i \to d$).

The single premium calculation can be simplified by means of an interesting approximation. Assume that the intensity of mortality for active lives is (approximately) equal to the intensity of disablement, thus:

$$\mu_y^{ad} = \mu_y^{ai} = \mu_y. \tag{3.191}$$

Then we have:

$$_t p_x^{\overline{aa}} = \exp\left[-\int_0^t (\mu_{x+u}^{ai} + \mu_{x+u}^{ad})\, du\right] = \exp\left[-\int_0^t 2\mu_{x+u}\, du\right] = {}_t p_{x,x} \tag{3.192}$$

where $_t p_{x,x}$ denotes the joint survival probability of a group of two lives both aged x, with the same mortality intensity μ_y. Denote by $\bar{A}_{x,x:\overline{n}|}$ the single premium of a joint life endowment assurance,

$$\bar{A}_{x,x:\overline{n}|} = \int_0^n {}_t p_{x,x} 2\mu_{x+t} v^t\, dt + {}_n p_{x,x} v^n. \tag{3.193}$$

Then, we can use the following approximation:

$$\bar{A}_{x,n}^{(\mathrm{E+PD})} \cong \bar{A}_{x,x:\overline{n}|}. \tag{3.194}$$

3.6 REFERENCES AND SUGGESTIONS FOR FURTHER READING

In actuarial calculations concerning disability benefits, the mathematics of Markov and semi-Markov chains provides both a powerful modelling tool and a unifying point of view, from which several calculation techniques and conventional procedures can be seen in a new light. To this purpose the reader is referred to Haberman (1988), Pitacco (1995a), Waters (1984, 1989), CMIR12 (1991).

Of course, much interesting actuarial work, both theoretical and practical, is not directly related to the Markov (or semi-Markov) multiple state approach. So, in this section we will provide some guidelines for further reading including also several contributions concerning traditional and/or approximate models for pricing and reserving in the field of disability benefits.

In a first category we collect contributions mainly describing (or in some sense related to) 'local' calculation methods, following the order adopted in sections 3.3 and 3.4.

The Danish model is briefly described by Ramlau-Hansen (1991). The new CMIB model, relating to practice in the UK, is presented and fully illustrated in CMIR12 (1991). The reader should consult also Sansom and Waters (1988), Waters (1984, 1989), Hertzman (1993), Cordeiro (1995). Rickayzen (1997) provides a preliminary discussion of the effect of policy lapses or terminations.

The Swedish model as well as relevant data are considered in many papers. The reader should consult Dillner (1969, 1974), Ekhult (1980), Källström (1990), Mattsson (1956), Mattsson and Unneryd (1968), Sandström (1990), Söderström (1980). An interesting comparative study is provided by Mattsson (1977). In Mattsson and Lundberg (1957) calculation techniques for sickness annuities and disabilities annuities are compared.

The method used in Norway (the so-called 'j-method') is described in particular by Sand (1968) and Sand and Riis (1980). The paper by Bjoraa (1951) describes the introduction of the j-method and hence is also of historical interest; data to implement this method are presented in Bjoraa (1960). Results from statistical investigation supporting the j-method are discussed by Ore, Sand and Trier (1964).

The Finnish model is described by Hännikäinen (1988), Kuikka, Lindqvist and Voivalin (1980) and Tuomikoski (1988).

The Inception/Annuity method adopted in the US is described in many textbooks. In particular, readers are referred to Jordan (1982) and Bowers *et al.* (1986). Readers interested in statistical data can consult, for example, Report of the Committee on Individual Health Insurance (1990).

European implementations of the Inception/Annuity method are described by Segerer (1993). In particular, the probabilistic structure and the statistical bases adopted in Switzerland are described by Chuard and Chuard (1992) and Chuard (1993); the use of transition intensities instead of probabilities to represent recoveries is discussed by Chuard M. (1995) and Chuard P. (1995).

The Dutch model is discussed by Gregorius (1993). Readers should consult also the textbook by Wolthuis (1994).

The traditional British 'Manchester Unity' method is described in the textbook by Neill (1977); related statistical aspects are dealt with by Benjamin and Pollard (1980). Readers are also referred to Bond (1963), Sanders and Silby (1988) and Turner (1988). A probabilistic critique of the quantities involved in actuarial calculations according to the 'Manchester Unity' method is presented by Haberman (1988). Statistical bases are in particular provided by CMIR7 (1984) and CMIR8 (1986).

Turning to 'country independent' studies, we can define a second category including papers and books dealing with general aspects of disability actuarial models. The reader interested in comparing different calculation techniques for disability annuities should consult Hamilton-Jones (1972), Mattsson (1977), Westwood (1972). A formal presentation of the actuarial theory underlying pricing and reserving for disability benefits is given by Türler (1970).

The handbooks by Swiss Re Group (1982) and Münchener Rück (1993) also include actuarial and insurance aspects of disability benefits. The textbook by Depoid (1967), which provides an extensive illustration of statistical and actuarial problems concerning several non-life insurance covers, also includes a detailed description of the time-discrete probabilistic structure of disability annuities. The textbook by Alegre (1990) contains a detailed analysis of actuarial problems related to disability benefits. Courant (1984) deals with general aspects of disability insurance (annuities as well as lump sums).

Very important examples of 'historical' actuarial work are offered by Hamza (1900), Du Pasquier (1912, 1913), Galbrun (1933), Mattsson (1930), Simonsen (1936), Stoltz (1930). The approximation problem consisting in replacing probabilities $_hp_x^{aa}$ by $_hp_x$ when calculating level annual premiums is analysed by Jacob (1934).

We conclude this section by listing some papers devoted to particular actuarial topics of disability benefits and related fields.

Several papers deal with probabilistic aspects of disability benefits and relevant data; see for example Lundberg (1969). In particular, the papers by Steffensen (1949, 1950) analyse the transition intensities involved in disability actuarial calculations. Seal (1970) studies the probability distributions of random time spent in sickness in any given interval allowing for recoveries.

As far as semi-Markov processes are concerned, the first application to disability benefits appears in Janssen (1966). The papers by Hoem (1972) and Hoem and Funck-Jensen (1982) deal with semi-Markov processes in a more general demographic framework.

Bull (1980) deals with actuarial problems, and in particular approximation methods, for evaluating disability lump sums. Disability group insurance is analysed by Hoem, Riis and Sand (1971) with special emphasis on single recurrent premiums; related problems are also discussed in Hoem (1988). Problems concerning the qualifying period in disability insurance are analysed by Hoem (1969b). Non-reported claims in disability insurance are studied by Waters (1992). Vaupel *et al.* (1979) provide an introductory discussion of modelling frailty and its effect on mortality dynamics.

4

The graduation of transition intensities

4.1 INTRODUCTION TO GRADUATION

Graduation may be regarded as the principles and methods by which a set of observed (or crude) probabilities are adjusted in order to provide a suitable basis for inferences to be drawn and further practical computations to be made. One of the principal applications of graduation is the construction of a survival model (usually presented in the form of a life table) or a multiple state model (as presented in Chapters 1–3). Graduation ensures that the resulting survival or multiple state model displays the required degree of smoothness and other desirable mathematical properties so that the functions calculated from the model for practical use (for example, premiums and reserves) also share these properties.

We consider for the moment a set of age specific 'crude' transition intensities $\overset{\circ}{\mu}_x^{ai}$, in the notation of Chapter 3, which have been calculated from a set of observations. These crude intensities, which are used in the multiple state model, can be regarded as a sample from a larger population and thus they contain some random fluctuations. If we believed that the true intensities, μ_x^{ai}, were mutually independent then the crude values would be our final estimates of the true, underlying intensities. However, a common, prior opinion about the form of these true intensities is that each is closely related to its neighbours. This relationship is expressed by the belief that the true intensities progress smoothly from one age to the next. So the next step is to graduate the crude intensities in order to produce smooth estimates, $\tilde{\mu}_x^{ai}$, of the true intensities. This is done by systematically revising the crude values, in order to remove the random fluctuations. (This can be thought of as a cheaper and more practicable alternative to increasing the size of the original investigation.)

Although this chapter refers only to the case of graduating transition intensities in the context of the disability insurance model of Chapter 3, the approach can be readily extended to probabilities rather than intensities and to the survival model (or life table) as well as to more general multiple state models and multiple decrement models, with intensities that depend on current age and a measure of duration (for example, length of current spell of disability).

There are many graduation methods that have been suggested in the literature, for example:

- graphic methods;
- parametric methods;
- non-parametric methods.

Our purpose is not to review the extensive literature on graduation, but we shall instead describe a flexible approach to parametric methods that has been successfully implemented in a wide variety of contexts. Interested readers should consult Benjamin and Pollard (1980) and London (1985) for a full presentation of this subject and Forfar *et al.* (1988) and Renshaw (1991) for a further discussion of parametric methods and Copas and Haberman (1983), London (1985) and Gavin *et al.* (1993, 1994, 1995) for a further discussion of non-parametric methods.

We note that parametric methods involve the fitting of some mathematical function (e.g. polynomial, spline or logistic function), with the parameters being determined by a formal procedure like maximum likelihood, minimum chi-square or weighted least squares. Although in the context of the assumed function, such methods will be efficient, they are liable to some degree of bias since no pre-assigned function will fit reality exactly.

Non-parametric methods aim to give more stable estimates than the crude values by combining data at different values of x, but without presupposing any particular mathematical form for, say, μ_x^{ai}. Like parametric methods, they are also liable to give biased estimates but in such a way that it is possible to balance (in some sense) an increase in bias with a decrease in sampling variation.

In this chapter, we shall present a parametric approach to graduation based on generalized linear models. Before discussing these methods, we first need to consider the nature of the crude estimates of the transition intensities.

4.2 CRUDE ESTIMATES OF THE TRANSITION INTENSITIES

We consider here the three-state model of Chapter 3 assuming the Markov property: extensions to the more general case are straightforward. We consider an observation period of perhaps several calendar years and assume

that each individual represents an independent realization of the underlying stochastic process $S(y)$, where y is the individual's age. We assume
that while under observation, we can observe the time and type of each
transition that an individual makes. We focus, for estimation purposes,
on an age interval, which we take as $(x, x+1)$ without loss of generality,
over which it is reasonable to assume that the transition intensities are
constants, $\mu_x^{ai}, \mu_x^{ia}, \mu_x^{ad}, \mu_x^{id}$.

The observations in respect of a single life are:

- the times between successive transitions;
- the numbers of transitions of each type.

With the assumption that the transition intensities are constant, equations
(1.38a) and (1.38b) show that each spell of length t ($0 \leq t \leq 1$) in the active
or invalid states contributes a factor of the form

$$\exp(-(\mu_x^{ad} + \mu_x^{ai})t)$$

or

$$\exp(-(\mu_x^{id} + \mu_x^{ia})t)$$

respectively to the likelihood so it suffices to record the total waiting time
spent in each state (as a 'sufficient statistic'). We then define

C_j = waiting time of the jth life in the active state

W_j = waiting time of the jth life in the invalid state

S_j = number of transitions from active to invalid by the jth life

R_j = number of transitions from invalid to active by the jth life

D_j = number of transitions from active to dead by the jth life

U_j = number of transitions from invalid to dead by the jth life

and define $C = \sum_1^N C_j$ (and so on), and use lower case letters for the
corresponding observed quantities. It can then be shown that the
likelihood for the four parameters given the observed data is

$$L(\mu_x^{ad}, \mu_x^{ai}, \mu_x^{id}, \mu_x^{ia}) = e^{-(\mu_x^{ad}+\mu_x^{ai})c} e^{-(\mu_x^{id}+\mu_x^{ia})w} (\mu_x^{ad})^d (\mu_x^{id})^u (\mu_x^{ai})^s (\mu_x^{ia})^r. \quad (4.1)$$

The waiting times, C and W, are usually referred to in the actuarial literature as the central exposed to risk.

The likelihood factorizes into functions of each parameter that are of
the form $e^{-\mu v} \mu^g$ so that the maximum likelihood estimators for the four
parameters are respectively

$$\hat{\mu}_x^{ad} = \frac{D}{C}, \quad \hat{\mu}_x^{id} = \frac{U}{W}, \quad \hat{\mu}_x^{ai} = \frac{S}{C}, \quad \hat{\mu}_x^{ia} = \frac{R}{W}. \quad (4.2)$$

We note that each estimator is the ratio of two random variables: number of transitions and waiting time (or central exposed to risk).

It may be important to be able to estimate the moments of these estimators, for example when comparing the results of two sets of observations or comparing one experience with a benchmark. It is a standard result of maximum likelihood theory that the asymptotic distribution of each $\hat{\mu}$ is normal with mean μ and variance $\mu/E(C)$ or $\mu/E(W)$ as appropriate. Sverdrup (1965) reports on further useful and elegant results, demonstrating that for each j

$$E(D_j) = \mu_x^{ad} E(C_j) \quad E(S_j) = \mu_x^{ai} E(C_j) \tag{4.3a}$$

$$E(U_j) = \mu_x^{id} E(W_j) \quad E(R_j) = \mu_x^{ia} E(W_j) \tag{4.3b}$$

$$E(D_j - \mu_x^{ad} C_j)^2 = E(D_j) \quad E(S_j - \mu_x^{ai} C_j)^2 = E(S_j) \tag{4.4a}$$

$$E(U_j - \mu_x^{id} W_j)^2 = E(U_j) \quad E(R_j - \mu_x^{ia} W_j)^2 = E(R_j) \tag{4.4b}$$

$$E(D_j - \mu_x^{ad} C_j)(S_j - \mu_x^{ai} C_j) = 0 \tag{4.5}$$

and so on, so that the quantities $D_j - \mu_x^{ad} C_j$, $S_j - \mu_x^{ai} C_j$, $U_j - \mu_x^{id} W_j$, and $R_j - \mu_x^{ia} W_j$ are uncorrelated.

Remark 1

We note that relationships (4.4a) and (4.4b) are analogous to the results for simple Poisson processes where $EX = \lambda T$ and $\operatorname{Var} X = E(X - \lambda T)^2 = EX$, where X is the number of events during a time interval of length T, with the change that here we are dealing with a case where T is a random variable. We also note that the form of the likelihood in equation (4.1) is identical to that which comes from assuming a Poisson distribution for the number of transitions conditional on the waiting times.

Remark 2

We note that $(D_j, U_j) = (0, 0)$, $(0, 1)$ or $(1, 0)$ and that, assuming that the jth life starts in the active state, $R_j = S_j$ or $S_j - 1$. Then, the estimators (4.2) are not independent but they are asymptotically independent. Thus the vector $(\hat{\mu}_x^{ad}, \hat{\mu}_x^{id}, \hat{\mu}_x^{ai}, \hat{\mu}_x^{ia})$ has an asymptotic multivariate normal distribution, and each component has a marginal asymptotic normal distribution with mean μ and variance $\mu/E(C)$ (or $\mu/E(W)$); and, asymptotically, the components of the vector are uncorrelated because of equation (4.5) and hence are independent (because of the normality property).

Remark 3

In the above discussion, we have presented a maximum likelihood approach to the estimation of the transition intensities. Other approaches have been suggested in the literature, principally in the context of mortality analysis: for example, the Product Limit Estimator due to Kaplan and Meier (1958), the Nelson–Aalen estimator (Nelson, 1972) and the so-called 'actuarial' estimator (Benjamin and

Pollard, 1980). The maximum likelihood approach has been chosen here because of the desirable properties that the resulting estimators enjoy, and because of the direct link to the parametric graduation methods that have been used in the literature and are illustrated in the subsequent sections. For a full discussion, readers are referred to Elandt-Johnson and Johnson (1980), London (1988) and Puzey (1997).

4.3 INTRODUCTION TO GENERALIZED LINEAR MODELS

4.3.1 Classical models

We shall begin this introduction to generalized linear models by briefly considering classical linear models.

Classical linear models comprise a vector of n independent normally distributed response random variables $\mathbf{Y}$ with means $m = E(\mathbf{Y})$ and constant variance σ^2. A non-random, systematic structure is incorporated by assuming the existence of a vector of covariates $\mathbf{x}_1, \mathbf{x}_2, \ldots, \mathbf{x}_p$ with known values such that:

$$\mathbf{m} = \sum_j \beta_j \mathbf{x}_j = \mathbf{X} \cdot \boldsymbol{\beta}$$

where $\mathbf{x}_j$ is an $n \times 1$ vector, $\boldsymbol{\beta}$ is a $p \times 1$ vector, $\mathbf{X}$ is an $n \times p$ matrix and $j = 1, 2, \ldots, p$.

So letting i index the observations, we have that:

$$E(Y_i) = m_i = \sum_{j=1}^{p} \beta_j x_{ij}$$

where x_{ij} is the value of the jth covariate for observation i.

The β_js are usually unknown parameters which have to be estimated from the response data $\mathbf{y}$ assumed to be a realization of $\mathbf{Y}$. Specific choices of the design matrix $\mathbf{X}$ lead to a broad class of linear models which includes the familiar regression type model in which Y_i are independent random variables distributed as $N(\alpha + \beta x_i, \sigma^2)$ and the respective one-factor and two-factor non-interactive models which form the basis of the familiar analysis of variance tests, namely:

Y_i are independent random variables distributed as $N(\alpha_i, \sigma^2)$
Y_{ij} are independent random variables distributed as $N(\alpha_i + \beta_j, \sigma^2)$.

Such models are traditionally fitted by a least squares method which, because of the independent normal assumption, is equivalent to maximum likelihood. An unbiased estimator based on the residual sum of squares is taken for σ^2. The adequacy of the model, including the strong constant variance assumption, normal assumption and systematic structure are monitored through residual plots. Attempts to simplify the systematic structure by nesting models may be tested statistically using the familiar F-test.

4.3.2 Generalizations

Classical linear models may be generalized in two respects: firstly, through the introduction of a much wider class of distributions, the so-called exponential family of distributions; and, secondly by linking the systematic component or linear predictor

$$\eta = \mathbf{X} \cdot \boldsymbol{\beta} \tag{4.6}$$

to the means **m** of the independent response variables through the introduction of a monotonic differentiable function g where

$$\eta = g(\mathbf{m}) \tag{4.7}$$

so that

$$\mathbf{m} = g^{-1}(\eta) = g^{-1}(\mathbf{X} \cdot \boldsymbol{\beta}).$$

The function g is called the **link function**.

The exponential family of distributions includes the normal, binomial, Poisson and gamma distributions amongst its members. Clearly, the normal distribution has to be selected in conjunction with the identity link function in order to retrieve the classical linear model (described above).

4.3.3 Exponential family

It will be helpful to consider some of the properties of the exponential family of distributions: these are described below in terms of a single observation, y.

A one-parametric exponential family of distributions has a log-likelihood of the form:

$$l = \frac{y\theta - b(\phi)}{\phi} + c(y, \phi)$$

where θ is the canonical parameter and ϕ is the dispersion parameter, assumed known. It is then straightforward to demonstrate that

$$m = E(Y) = \frac{\mathrm{d}}{\mathrm{d}\theta} b(\theta), \quad \mathrm{Var}(Y) = \phi \frac{\mathrm{d}^2}{\mathrm{d}\theta^2} b(\theta) = \phi b''(\theta).$$

We note that $\mathrm{Var}(Y)$ is the product of two quantities. $b''(\theta)$ is called the variance function and depends on the canonical parameter and hence on the mean. We can write this as $V(m)$.

By way of illustration, the log-likelihoods for some common distributions are given below.

Normal

$$l = \frac{my - \frac{1}{2}m^2}{\sigma^2} - \frac{y^2}{2\sigma^2} - \frac{1}{2}\log(2\pi\sigma^2)$$

$$\theta = m, \quad b(\theta) = \frac{\theta^2}{2}, \quad V(m) = 1, \quad \phi = \sigma^2.$$

Poisson

$$l = y\log m - m - \log(y!)$$

$$\theta = \log m, \quad b(\theta) = e^\theta, \quad V(m) = m, \quad \phi = 1.$$

Binomial

Suppose $D \sim$ binomial (N, m). Define $Y = D/N$. Then:

$$l = \frac{y\log[m/(1-m)] - \log(1-m)}{1/N} + \log\binom{N}{Ny}.$$

Hence $\quad \theta = \log[m/(1-m)], \quad b(\theta) = \log(1+e^\theta), \quad V(m) = m(1-m),$
$\phi = 1/N$.

Gamma

Suppose $Y \sim$ gamma with mean m and variance m^2/ν:

$$l = \frac{-(y/m) + \log(1/m)}{1/\nu} + \nu\log y + \nu\log\nu - \log\Gamma(\nu)$$

$$\theta = \frac{-1}{m}, \quad b(\theta) = -\log(-\theta), \quad V(m) = m^2, \quad \phi = \nu^{-1}.$$

4.3.4 Implementation

The GLIM computer package (like other modern statistical packages) is designed to enable the user to fit generalized linear models interactively. It offers the choice of modelling distribution, link function and linear predictor. Parameter estimates, fitted values and a goodness of fit measure, called the deviance, all form part of the output as each model is fitted; while a variety of residual plots can be displayed. In addition, users may add their own programs (or macros) to GLIM, thereby providing further versatility.

Estimation of the linear predictor parameters β is by maximum likelihood using an iterative weighted least squares algorithm. There exist sufficient statistics for the parameters in the linear predictor (equation (4.6)) provided that the modelling distribution is matched with a specific link function called the **canonical** or **natural** link function, given by $\eta = \theta(\mathbf{m})$. Alternative link functions are, however, available. Examples of the canonical link functions are (see above):

- Normal: identity;
- Poisson: log;
- Binomial: logit;
- Gamma: reciprocal.

With n observations $\mathbf{y}$ available, models with between 1 and n parameters may be considered. The extreme cases have an important role to play. The null model, comprising a single parameter, has the property that all the $\mathbf{m}$ components are identical, leaving all of the variation in the data $\mathbf{y}$ to be accounted for in the error structure of the model. At the other extreme, with n independent data components and n parameters to estimate, the so-called saturated model is defined in which the fitted values are the data themselves ($\hat{\mathbf{m}} = \mathbf{y}$). Clearly, we seek an optimum model somewhere between these extremes. Ideally, it should involve as few parameters as possible while accounting for the salient structure present in the response data, leaving a pattern-free set of residuals.

The examination of residual plots plays an important role in the assessment of the viability of any proposed model.

Goodness of fit is based on the likelihood ratio principle with the saturated model providing the benchmark, rather than on an adaptation of the possibly more familiar Pearson (chi-square) goodness of fit criterion. Consider any intermediate model, with p parameters, called the **current** model. Minus twice the logarithm of this ration (a monotonic mapping) is defined to be the (current) model deviance. Nested models can then be compared on the basis of the differences between their model deviances. In the case of the classical linear model, the model deviance reduces to the familiar residual sum of squares term.

It is important that inferences should be based on differences between the deviances for different models since their absolute values are conditional on the total number of covariates under simultaneous investigation. Differences between the deviances may be referred to the chi-square distribution with appropriate degrees of freedom for a formal test (this is an approximate result). No attempt should be made to interpret the absolute values of the model deviance statistic.

The reader is referred to McCullagh and Nelder (1989) for a more detailed description as well as to the GLIM manual: Francis *et al.* (1993).

4.4 GRADUATION OF TRANSITION INTENSITIES USING GENERALIZED LINEAR MODELS

4.4.1 Introduction

In this section, we present a comprehensive methodology for graduating the transition intensities in a three-state multiple state model set up for application to permanent health insurance. The approach is based on generalized linear models (GLMs) and is exemplified by using the data collected and analysed by the UK Continuous Mortality Investigation (CMI) Bureau (under the auspices of the Institute and Faculty of Actuaries) in respect of the male standard experience for individual PHI policies for the period 1975–78, as a case study. The comprehensive and versatile nature of this approach is shown to be applicable to three sets of transition intensities for which data are available: for sickness recovery (as functions of age at sickness onset, x, and duration of sickness, z), for death as sick (also a bivariate function of x and z) and for sickness inception (a function of x only). The full potential of the GLM methodology means that approximations to normality which would lead to complex, iterative methods of fitting can be avoided and also that the presence of duplicate policies can be allowed for. Full details of the approach are presented by Renshaw and Haberman (1995).

The underlying process used for modelling permanent health insurance (PHI) business has been presented in Fig. 3.1. As we have indicated, a detailed knowledge of the four transition intensities, μ^{ai}, μ^{ad}, μ^{ia}, μ^{id}, is essential for the calculation of the premiums, reserves and annuity values needed for the actuarial management of this business. Under the terms of PHI business, benefits become payable if state i continues to be occupied at the end of an agreed deferred period, f (as described in section 3.2), which typically is either 1, 4, 13, 26 or 52 weeks, and which commences immediately upon entry into state i (from state a). CMIR12 (1991) provides a full description of the permanent health insurance data collected by the CMI Bureau from UK life insurance companies.

To simplify the presentation in this section, we will temporarily modify our standard notation, replacing μ^{ai}, μ^{ad}, μ^{ia} and μ^{id} by σ, μ, ρ and ν respectively.

There are certain features which arise from the mode of data collection and the design of individual PHI policies which create difficulties for the estimation of the underlying parameters of this model. Thus, problems occur because:

1. duplicate policies are present;
2. transitions out of state i are observed only when the duration of the sickness spell is greater than the deferred period of the policy, f;

3. transitions from state a to state i are observed only when the duration of the subsequent sickness spell is greater than the deferred period of the policy, f;
4. transitions from state a to state d are not observed.

So the data available relating to PHI policies permit the graduation of three of the four transition intensities namely σ, ρ, ν, subject to these difficulties being resolved.

The detailed results presented here are for data collected in the four-year observation window beginning 1975 to the end of 1978. The data used to graduate each transition intensity ρ, ν or σ, consists of a set of ordered pairs (i_u, e_u) defined over a network of cells or units denoted by u. Here i_u and e_u are the number of transitions and corresponding central exposures recorded for the four year observation window. The precise definition of i_u and e_u and the nature of the units u will be given in sections 4.4.2 to 4.4.4 which deal separately with the graduation of ρ, ν and σ respectively. Graduations are constructed using the generalized linear modelling (GLM) framework based on independent Poisson response variables i_u. Not only does such a framework offer a wide choice of para-meterized graduation formulae, it also provides the means of fitting and assessing the choice of formula. The justification for choosing Poisson responses comes from the assumption that a transition intensity (ρ, ν and σ) is constant within each cell or unit. Then, as noted in section 4.2, the form of the likelihood function is identical to that which comes from assuming a Poisson distribution for the number of transitions: see Sverdrup (1965) for the case of the multiple state model and Forfar *et al.* (1988) for the case of estimating the force of mortality.

For deferred period $f = 52$ weeks, the aggregate exposed to risk was 40,981 days and there were nine recoveries and five deaths observed. Because of the paucity of data in this category, data from this deferred period have been omitted from the analyses reported in the following sections.

4.4.2 On the graduation of the sickness recovery intensities ρ

The data

Individual data sets are available for each deferred period $f = 1, 4, 13$ or 26 weeks. For each f denote these $(i_u, e_u) \equiv (i_{x,z}, e_{x,z})$ where i_u are the recorded number of sickness recoveries from exposures e_u defined over a rectangular grid of cells or units $u \equiv (x, z)$. This is defined by grouping ages x into eight categories 20–, 30–, 35–, 40–, 45–, 50–, 55–, 60–64 and sickness duration z into individual weekly categories up to week 30, then grouping weeks 30–, 39–52, followed by years 1–, 2–, 3–, 4–, 5–11. For modelling purposes, the data are located at the centroid of their respective cells determined by weighted averages, with the relative exposures as the weights.

In utilizing these data, we note that a few claim revivals are present in the records and that these have been treated as 'negative' recoveries and their numbers 'netted off' against the number of observed recoveries.

The graduation process

Sickness recovery intensities $\rho_{x,z}$ are regarded as functions of two covariates, age x at the onset of sickness and sickness duration z. The graduation process is designed to establish the nature of the smooth functional relationship and is here treated as an exercise in surface fitting. In the GLM framework introduced in section 4.3, the class of functions available takes the form

$$g(\rho_{x,z}) = \eta_{x,z}$$

where g and η denote functions. The function g is both differentiable and one-for-one so that, in part, the inverse g^{-1} exists and

$$\rho_{x,z} = g^{-1}(\eta_{x,z}).$$

The function η is parameterized in order to introduce flexibility and takes the form of a linear predictor

$$\eta_{x,z} = \sum_j h_j(x,z)\beta_j$$

where the h_j define the known covariate structure of the current model under consideration. The predictor $\eta_{x,z}$ is linear in the unknown parameters β_j, although it is possible to relax this condition if necessary.

Estimates for the β_js and diagnostic checks are needed to implement the graduation process. These are based on the Poisson modelling assumption

$$i_u \sim \text{Poi}(e_u\rho_u) \text{ independently for all } u \equiv (x,z)$$

with respective mean and variance

$$m_u = \text{E}(i_u) = e_u\rho_u, \quad \text{Var}(i_u) = m_u.$$

Expressed in terms of the mean $\mathbf{m} = (m_u)$, the associated log-likelihood is

$$\log(l(\mathbf{i};\mathbf{m})) = \sum_u \{-m_u + i_u \log(m_u)\} + \text{constant}$$

where $\mathbf{i} = (i_u)$. The unknown β_js enter the log-likelihood by substitution via the relationship

$$m_u = e_u g^{-1}(\eta_u) = e_u g^{-1}\left(\sum_j h_j(u)\beta_j\right)$$

and their values, $\hat{\beta}_j$ are estimated by maximizing the log-likelihood. Denoting the resulting fitted values by

$$\hat{m}_u = e_u g^{-1}\left(\sum_j h_j(u)\hat{\beta}_j\right),$$

the optimum value of the log-likelihood under the current model (predictor) structure, c, is

$$\log(l_c) = \log(l(\mathbf{i}; \hat{\mathbf{m}})) = \sum_u \{-\hat{m}_u + i_u \log(\hat{m}_u)\} + \text{constant}.$$

The constant is evaluated by reference to the saturated model structure, s, which has the property that its fitted values $\hat{m}_u = i_u$, the responses, constitute a perfect fit. Under the saturated model, the value of the log-likelihood becomes

$$\log(l_s) = \log(l(\mathbf{i}; \mathbf{i})) = \sum_u \{-i_u + i_u \log(i_u)\} + \text{constant}$$

giving rise to the model deviance

$$D(c, s) = \sum_u d_u = -2\log(l_c/l_s) = -2\log(l_c) + 2\log(l_s)$$

$$= 2\sum_u \{-(i_u - \hat{m}_u) + i_u \log(i_u/\hat{m}_u)\}.$$

As in Renshaw (1991), a range of statistical and graphical diagnostic checks is used for measuring the adequacy of any model in representing the crude data.

The overall measure of goodness of fit is provided by the model deviance, as defined above. It is possible to compare the goodness of fit of the various hierarchical model predictor structures fitted to the same data by differencing the resulting model deviances. Asymptotically, these differences have the chi-square distribution. This approach is used to identify the most appropriate model structure or graduation.

Deviance residuals defined by

$$r_u = \text{sign}(i_u - \hat{m}_u)\sqrt{d_u},$$

where d_u denote the individual components of the model deviance, are used extensively to monitor the 'quality' of the fit. Plots of deviance residuals against age, sickness duration and fitted (expected) values and histograms of deviance residuals are widely used as visual checks of the underlying modelling assumptions. The reader is referred to Chapter 10 of McCullagh and Nelder (1989) for a further commentary on diagnostic model checking.

Two specific link functions g are of particular interest in this context, the log-link and the parameterized power-link. We restrict our discussion to the former in this analysis. To implement the log-link for which

$$\log(m_u) = \log(e_u) + \log(\rho_u),$$

the $\log(e_u)$ terms are declared as offsets and the graduated recovery transition intensities are

$$\hat{\rho}_{x,z} = \exp\left\{\sum_j h_j(x,z)\hat{\beta}_j\right\}.$$

Two-factor additive structure under the log-link

Under this heavily parameterized structure, in which both the age categories x and sickness duration categories z are treated as factors, the linear predictor is assigned the additive structure

$$\eta_{x,z} = \mu + \alpha_x + \beta_z$$

so that under the log-link the graduation formula is

$$\rho_{x,z} = \exp(\mu + \alpha_x + \beta_z).$$

It is not proposed to pursue this graduation formula further, since it is excessively parameterized and lacks interaction terms which are found to be necessary for the successful graduation of these data. However, an exploratory analysis indicates that, apart from some fluctuations for low values of z, $\hat{\beta}_z$ appears to reduce linearly with $\sqrt{z}$ and that $\hat{\alpha}_x$ appears to reduce with increasing x.

Graduation for deferred period one week

Given the above comments, we approach the graduation for deferred period one week using the log-link and a range of possible linear predictors based on the two variates x and $\sqrt{z}$. After extensive testing, the graduation formula

$$\rho_{x,z} = \exp(\beta_0 + \beta_1 x + \beta_2 z + \beta_3 \sqrt{z} + \beta_4 xz + \beta_5 x\sqrt{z}) \qquad (4.8)$$

comprising the log-link in combination with a polynomial predictor in the variates x and $\sqrt{z}$ has been adopted. For a general discussion of regression using fractional polynomial predictors the reader is referred to Royston and Altman (1994). The parameter estimates, and their standard errors, are reproduced in Table 4.1. Diagnostic checks based on the deviance residuals are highly supportive of the formula. A sample of graduated values is presented in Table 4.2.

As we will see from the following sub-section, the graduation for deferred period one week has a somewhat different structure compared to the longer deferred periods. This is not unexpected as this class of

Table 4.1 Parameter estimates, standard errors (s.e.) ρ-graduation, deferred period one week

	Estimate	s.e.
β_0	6.0006	0.1567
β_1	−0.04076	0.003469
β_2	0.05844	0.01050
β_3	−1.1546	0.1033
β_4	−0.0008759	0.0002436
β_5	0.007937	0.002261

business was at that time sold by a small group of companies only and was designed for a narrow group of occupations. In more recent years, little new business has been sold with this deferred period (Clark and Dullaway, 1995).

Graduations for deferred periods greater than one week

One of the notable features of the sickness recovery data for deferred periods f in excess of one week is the low recovery rates associated with the weeks immediately after the sickness benefit becomes payable.

Table 4.2 Graduated sickness recovery intensities, deferred period one week

Age		20	30	40	50	60
	1	68.7488	49.0790	35.0370	25.0125	17.8562
	2	47.4132	34.6740	25.3576	18.5444	13.5618
	3	35.9926	26.7590	19.8943	14.7906	10.9963
	4	28.7147	21.6169	16.2735	12.2510	9.2228
	5	23.6474	17.9807	13.6719	10.3957	7.9045
D	6	19.9185	15.2697	11.7060	8.9739	6.8795
U	7	17.0667	13.1731	10.1677	7.8481	6.0576
R	8	14.8225	11.5067	8.9327	6.9344	5.3832
A	9	13.0164	10.1539	7.9209	6.1789	4.8201
T	10	11.5368	9.0368	7.0786	5.5447	4.3432
I	15	6.9757	5.5334	4.3894	3.4819	2.7620
O	20	4.7134	3.7530	2.9883	2.3794	1.8946
N	30	2.6085	2.0608	1.6281	1.2863	1.0162
	40	1.6892	1.3077	1.0123	0.7837	0.6067
	50	1.2094	0.9101	0.6848	0.5154	0.3878
	52.2	1.1359	0.8489	0.6343	0.4740	0.3542
	104.4	0.4884	0.2930	0.1758	0.1055	0.0633
	156.5	0.4200	0.1915	0.0873	0.0398	0.0181
(in weeks)	208.7	0.5172	0.1741	0.0586	0.0197	0.0066
	260.9	0.8010	0.1954	0.0477	0.0116	0.0028

There is evidence from other investigations to indicate that this is a real feature (Medin (1951), Mattson (1956) and Dillner (1969) in respect of Swedish experience). A number of commentators have agreed that a possible reason for this 'run-in period' of about four weeks is the differential attitude to claiming among a particular group of policyholders. Thus, those near to recovery at the end of the deferred period for their policy seem to be less likely to bother to submit a claim, on the grounds that the claim would be short-lived. Here, the corresponding (imminent) recoveries are excluded from the observed data.

CMIR12 (1991) have modelled this effect in an *ad hoc* manner by including as an adjustment term a multiplicative factor which itself takes a linear form. Our approach is more general and we have been able to model this effect successfully by employing break-point predictor terms (similar to natural splines) of the type

$$\sum_{j=0}^{J} \beta_{0j} z^j + \sum_{k=1}^{K} \sum_{j=1}^{J} \beta_{kj} (z - z_{k+})^j$$

with knots z_k where $(z - z_k)_+ = z - z_k$ if $z > z_k$, and $(z - z_k)_+ = 0$ otherwise. In practice, two knots z_1, z_2 in combination with line segments $(J = 1)$ are found to be optimal. Thus the graduation formula

$$\rho_{x.z} = \exp(\beta_0 + \beta_1 x + \beta_2 z + \beta_3 (z - z_1)_+ + \beta_4 (z - z_2)_+$$
$$+ \beta_5 x (z - z_1)_+) \tag{4.9}$$

has been adopted to graduate the data for deferred periods of four and 13 weeks separately. The knots are located by constructing a deviance profile through scanning the possible positional choices for the pair of knots. These deviance profiles are reproduced in Table 4.3; locating the pair of knots at 6.5 and 45.5 weeks for deferred period four weeks and at 16.5 and 45.5 weeks for deferred period 13 weeks. We would expect that the first knot would correspond with the four-week 'run-in period' and that $\hat{\beta}_3 < 0$. Another feature of the observations is the change of gradient of ρ with respect to z for durations of over one year (i.e. a reduction in the steepness). So we would expect that the second knot would be positioned near to 52 weeks duration and that $\hat{\beta}_4 > 0$. The detailed parameter estimates, and their standard errors, for the two deferred periods are reproduced in Table 4.4 and samples of the resulting graduations are given in Tables 4.5 and 4.6. The associated deviance residual plots are highly supportive of the graduations.

Perhaps the effect of the break-point terms is most dramatically illustrated in the deviance residual plots against sickness duration reproduced in Figs 4.1 and 4.2 for deferred period four weeks. Figure 4.1 comes from an attempt to graduate sickness duration effects as a cubic in z. This approach is clearly too rigid and dramatically fails to model the low

Table 4.3 ρ-graduations. Deviance profiles for positioning of two knots

z_2	z_1 5.5	6.5	7.5	8.5	9.5	z_2	z_1 14.5	15.5	16.5	17.5	18.5
13.5	329.7	309.6	304.6	307.4	315.9						
15.5	323.7	302.1	297.1	300.4	308.8						
17.5	318.3	296.3	291.9	295.8	304.6						
19.5	314.5	292.5	289.0	293.8	303.1						
21.5	311.6	290.6	288.0	293.6	303.4	21.5	226.4	223.0	216.6	221.3	224.2
23.5	309.3	289.1	287.3	293.5	303.6	23.5	223.4	219.8	214.1	218.9	221.9
25.5	307.3	287.6	286.5	293.2	303.5	25.5	220.3	216.5	211.1	215.9	218.9
27.5	305.5	286.4	285.9	293.0	303.5	27.5	217.9	214.2	209.3	214.3	217.4
29.5	303.4	284.8	284.6	292.0	302.6	29.5	214.7	210.8	206.1	211.0	214.1
34.5	299.2	281.8	282.8	290.9	301.7	34.5	208.2	204.6	200.5	205.6	208.7
45.5	295.1	280.1	282.8	291.8	302.7	45.5	196.8	193.8	190.5	195.7	199.0
78.5	321.4	314.3	320.9	331.0	340.9	78.5	202.3	202.5	202.2	207.5	210.7
Deferred period 4 weeks						Deferred period 13 weeks					

values of the recovery rates associated with the period immediately after the sickness benefit becomes payable. Figure 4.2 comes from including the break-point predictor.

Attempts to graduate the data for deferred period 26 weeks using formula 4.9 are less successful due to the sparseness of the data. The best that can be achieved to preserve the general pattern of graduations established in Tables 4.5 and 4.6 for deferred periods four and 13 weeks is by locating the pair of knots at 29.5 and 45.5 weeks while omitting the interaction term which is non-significant. Even so, it is difficult to justify the positioning of a knot at 29.5 weeks as is evident from Table 4.7, which gives the detailed parameter estimates and their standard

Table 4.4 Parameter estimates, standard errors (s.e.) ρ-graduation, deferred periods four and 13 weeks

	4-week deferred period		13-week deferred period	
	estimate	*s.e.*	*estimate*	*s.e.*
β_0	0.5617	0.3252	−3.042	1.343
β_1	−0.01394	0.002991	−0.02661	0.005914
β_2	0.3141	0.04870	0.3303	0.08234
β_3	−0.3390	0.05062	−0.3806	0.08492
β_4	0.04254	0.005140	0.5058	0.007395
β_5	−0.0007846	0.0001754	−0.0003316	0.0001379

Table 4.5 Graduated sickness recovery intensities, deferred period four weeks

Age		20	30	40	50	60
	4	4.66203	4.05554	3.52794	3.06899	2.66974
	5	6.38265	5.55232	4.83001	4.20166	3.65506
	6	8.73831	7.60152	6.61262	5.75237	5.00403
	7	10.0191	8.68161	7.52263	6.51837	5.64818
D	8	9.62070	8.27129	7.11108	6.11361	5.25605
U	9	9.23829	7.88037	6.72204	5.73398	4.89115
R	10	8.87099	7.50792	6.35429	5.37792	4.55158
A	15	7.24227	5.89367	4.79619	3.90308	3.17627
T	20	5.91258	4.62649	3.62014	2.83269	2.21653
I	30	3.94079	2.85091	2.06245	1.49205	1.07940
O	40	2.62656	1.75677	1.17501	0.78590	0.52565
N	50	2.12000	1.31096	0.81067	0.50130	0.30999
	52.2	2.12914	1.29428	0.78678	0.47828	0.29074
	104.4	2.35996	0.95265	0.38456	0.15524	0.06266
	156.5	2.61580	0.70120	0.18796	0.05039	0.01351
	208.7	2.89938	0.51611	0.09187	0.01635	0.00291
(in weeks)	260.9	3.21370	0.37988	0.04490	0.00531	0.00063

errors. With the following numbers of reported transitions for deferred period 26 weeks:

	Weeks						Years				
Sickness duration	26	27	28	29	30–	39–	1–	2–	3–	4–	5–11
No. of transitions	7	6	2	8	38	27	29	4	3	3	5

Table 4.6 Graduated sickness recovery intensities, deferred period 13 weeks

Age		20	30	40	50	60
	13	2.0547	1.5748	1.2067	0.9248	0.7087
D	14	2.8592	2.1911	1.6791	1.2868	0.9861
U	15	3.9785	3.0488	2.3364	1.7905	1.3721
R	20	5.3516	4.0537	3.0707	2.3250	1.7619
A	30	3.0304	2.2207	1.6273	1.1924	0.8738
T	40	1.7160	1.2165	0.8623	0.6113	0.4333
I	50	1.2201	0.8367	0.5738	0.3935	0.2698
O	52.2	1.2036	0.8194	0.5579	0.3798	0.2586
N	104.4	0.8670	0.4965	0.2843	0.1628	0.0932
	156.6	0.6246	0.3008	0.1449	0.0698	0.0336
	208.7	0.4490	0.1823	0.0739	0.0299	0.0121
(in weeks)	260.9	0.3241	0.1105	0.0377	0.0128	0.0044

Residuals vs. sickness duration (restricted region)

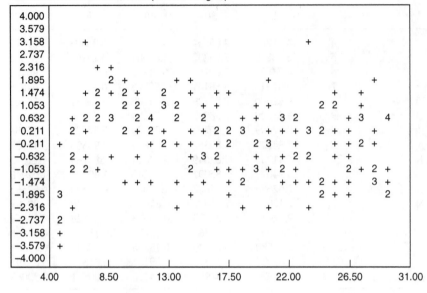

Figure 4.1 Log-link, polynomial predictor, deferred period four weeks.

clearly the data are too sparse to support a significant graduation pattern similar to those for deferred periods four and 13 weeks. Alternative graduation methods may be needed to achieve more satisfactory results: this has not been followed up here.

Residuals vs. sickness duration (restricted region)

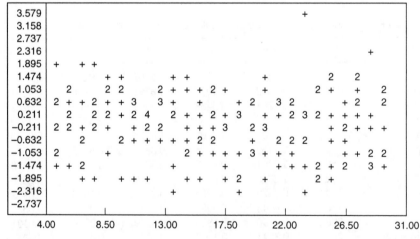

Figure 4.2 Log-link, break-point predictor, deferred period four weeks.

Table 4.7 Parameter estimates, standard errors (s.e.) ρ-graduation deferred period 26 weeks

	Estimate	s.e.
β_0	0.4341	4.074
β_1	−0.05949	0.007825
β_2	0.08571	0.1415
β_3	−0.1636	0.1511
β_4	0.06749	0.01830

4.4.3 On the graduation of the invalid to death intensities ν

The data and graduation process

The observed incidence of transitions from invalid to death is relatively low so that it is only meaningful to attempt to graduate the combined data set for the different deferred periods (i.e. 1, 4, 13 and 26 weeks). It has further been decided to work with a reduced rectangular grid of cells $u \equiv (x, z)$ defined by grouping age x into just six intervals 20–, 40–, 45–, 50–, 55–, 60–64 years and sickness duration z into just 14 intervals 1–, 4–, 7–, 10–, 13–, 16–, 19–, 22–, 26–, 30, 39–52 weeks, 1–, 2–, 5–11 years. Again the data are located at the centroid of their respective cells determined by weighted averages. The grouped data comprising the recorded numbers of deaths from sickness $i_{x,z}$ from exposures $e_{x,z}$ are presented in Table 4.8 in order to convey the orders of magnitude involved. The graduation process outlined in section 4.4.2 based on Poisson response variables $i_u \sim \text{Poi}(e_u \nu_u)$ independently for all $u \equiv (x, z)$ with mean and variance $m_u = E(i_u) = e_u \nu_u$, $\text{Var}(i_u) = m_u$ is again used here.

Graduation for the combined deferred periods

The graduation formula for the force of mortality

$$\nu_{x,z} = \exp(\beta_0 + \beta_1 x + \beta_2 z + \beta_3 (z - z_1)_+ + \beta_4 (z - z_2)_+) \qquad (4.10)$$

with two knots at sickness duration $z_1 = 14.5$ weeks and $z_2 = 182$ weeks (i.e. 3.5 years) based on the log link was adopted. The positions of the knots have again been located by constructing a deviance profile which is presented in Table 4.9. The parameter estimates and their standard errors are reproduced in Table 4.10, while interaction terms are not found to be statistically significant. The associated deviance residual plots (not shown here) display a degree of skewness consistent with Poisson responses of low order of magnitude, such as is the case here. A sample of the resulting graduations is reproduced in Table 4.11.

The formula is thus multiplicative in the two factors, age x and sickness duration z, with the age effect modelled as a Gompertz term. The locations

Table 4.8 Deaths and exposures by sickness duration and age, sickness to death transitions

Age	20–	40–	45–	50–	55–	60–64
Duration (in weeks)						
1–	2	0	1	7	3	6
	21724	8548	10672	14152	11500	11262
4–	1	0	3	3	4	3
	14223	7079	9135	11143	9420	8743
7–	2	1	4	2	5	2
	8384	4438	5817	7281	6404	6024
10–	1	0	2	1	3	2
	5405	3004	4069	5017	4783	4622
13–	4	2	2	3	3	3
	6384	3820	4716	5344	5316	4619
16–	0	2	1	1	0	2
	5157	3087	3814	4340	4574	3957
19–	1	3	0	2	7	1
	4311	2573	3264	3652	4074	3705
22–	1	0	1	0	0	2
	4600	2744	3878	4163	4694	4627
26–	0	3	3	0	3	3
	5187	3169	4153	5208	5687	5411
30–	1	4	3	5	4	5
	8935	6080	8201	10447	11539	10860
39–	3	1	5	4	3	1
	9862	8042	8976	12985	15802	13822
(in years)						
1–	6	4	3	8	11	7
	23180	22526	24545	39763	58076	39904
2–	1	2	1	4	12	5
	26709	37500	35559	67204	99152	40872
5–11	1	3	6	5	4	0
	17964	30121	32522	53347	36948	3

Table 4.9 ν-graduations; deviance profiles for positioning of two knots, all deferred periods

	z_1				
z_2	8.5	11.5	14.5	17.5	20.5
28	112.83	111.69	110.05	111.65	110.42
34.5	111.00	109.82	108.41	109.85	109.13
45.5	106.06	104.50	102.88	103.96	103.07
78	99.70	98.76	97.94	99.30	99.28
182	89.87	89.83	89.80	91.30	91.77
416	118.02	118.68	119.10	119.79	120.01

Table 4.10 Parameter estimates, standard errors (s.e.) ν-graduation, all deferred periods

	Estimate	*s.e.*
β_0	−3.549	0.4249
β_1	0.02125	0.007474
β_2	0.04156	0.01829
β_3	−0.05297	0.01898
β_4	0.01467	0.002844

Table 4.11 Sickness to death intensities; graduated values

Age		*20*	*30*	*40*	*50*	*60*
	1	0.04471	0.05529	0.06838	0.08456	0.10458
	2	0.04660	0.05764	0.07128	0.08815	0.10902
	3	0.04858	0.06008	0.07430	0.09189	0.11365
	4	0.05064	0.06263	0.07746	0.09579	0.11847
	5	0.05279	0.06529	0.08074	0.09986	0.12350
D	6	0.05503	0.06806	0.08417	0.10409	0.12874
U	7	0.05737	0.07095	0.08774	0.10851	0.13420
R	8	0.05980	0.07396	0.09147	0.11312	0.13989
A	9	0.06234	0.07710	0.09535	0.11792	0.14583
T	10	0.06499	0.08037	0.09939	0.12292	0.15202
I	15	0.07790	0.09635	0.11915	0.14736	0.18224
O	20	0.07359	0.09101	0.11255	0.13919	0.17214
N	30	0.06566	0.08120	0.10042	0.12419	0.15359
	40	0.05858	0.07245	0.08959	0.11080	0.13703
	50	0.05227	0.06464	0.07994	0.09886	0.12226
	52.2	0.05098	0.06305	0.07798	0.09643	0.11926
	104.4	0.02812	0.03477	0.04300	0.05318	0.06577
	156.5	0.01551	0.01918	0.02372	0.02933	0.03628
	208.7	0.01266	0.01565	0.01936	0.02394	0.02960
(in weeks)	260.9	0.01500	0.01855	0.02295	0.02838	0.03510

of the two knots represent turning points in the progression of ν with z. Thus, the presence of z_1 creates a peak in the values while z_2 causes the trend with duration to begin to increase again. These positions have been determined from the data via the deviance profile (as in Section 4.4.2).

4.4.4 On the graduation of the sickness inception intensities σ

Setting up the graduation process

Unlike the earlier two cases, the sickness inception intensity σ_x is a function of just one covariate, age x at onset of sickness. Thus $u \equiv x$ and ideally data (i_x, e_x) comprising the number of transitions from active to invalid with central exposures e_x over a range of ages x are required to graduate σ. Transition counts from active to invalid are not, however, available for analysis due to the left censoring induced by the deferred period f. Thus, sickness inception events go unrecorded if the time in state i (a) does not exceed f or (b) exceeds f but is not registered as a claim, for whatever reason.

We consider the nominal number of claims i_x, adjusted for non-registrations where necessary. Sickness inceptions may contribute more than once to these claim totals as they are based on policy counts, thereby inducing over-dispersion in the GLM as described by Renshaw (1992); while the nature of the adjustment for non-registration is discussed briefly below. Let $\pi_{x,f}$ denote the probability that the time in state i exceeds f, then the i_x are modelled as a GLM based on independent over-dispersed Poisson responses with

$$m_x = E(i_x) = \pi_{x,f} e_x \sigma_x \quad \text{and} \quad \text{Var}(i_x) = \phi_x V(m_x) = \phi_x m_x.$$

The dispersion parameters ϕ_x are needed to form weights $1/\phi_x$ in the model fitting process. The evaluation of both the central exposures e_x and the probabilities $\pi_{x,f}$ is described below. To implement the log-link for which

$$\log(m_x) = \log(\pi_{x,d}) + \log(e_x) + \log(\sigma_x),$$

the $\log(\pi_{x,d}) + \log(e_x)$ terms are declared as offsets in the model and the graduated values are then

$$\hat{\sigma}_x = \exp\left\{\sum_j h_j(x)\hat{\beta}_j\right\}.$$

Detailed graduations based on these methods are presented in a subsequent paragraph 4.4.4.

The data inputs

The number of claim inceptions i_x is taken as

$$i_x = ir_x + in_x$$

where ir_x are the reported claims and in_x represent an adjustment for incurred but not reported claims (allowing for the presence of duplicates). Further details are given in CMIR12 (1991) and in Renshaw and Haberman (1993).

The exposures e_x needed in the graduation process are not immediately available (from the data). They are estimated by eh_x in the following equation

$$eh_x = te_x - (cl_x + \bar{cl}_x)/52.18 \tag{4.11}$$

where eh_x, cl_x and $\bar{cl}_x$ denote the exposures spent in the four-year observation window at age x last birthday as active, invalid and claiming, invalid and not claiming respectively; the factor 52.18 reflecting the fact that the data on sickness exposures are given in weeks rather than years. The values of the total exposure, te_x, together with the values of cl_x and $\bar{cl}_x$ are estimated from the data. For te_x, adjustments are needed to allow for the durational effect and to ensure that new entrants do not contribute to an exposure until they have had their policies in force for at least the particular deferred period being considered. The values of cl_x are available directly from the observed data. For $\bar{cl}_x$, an estimation procedure based on $\pi_{x,t} = P(T_x > t)$ is used where T_x is the randomtime spent in the invalid state I induced by a sickness inception event occurring at age x years last birthday. We note that

$$\pi_{x,t} = \exp\left(-\int_0^{t/52} (\rho_{x+u,u} + \nu_{x+u,u})\, du\right). \tag{4.12}$$

Thus $\pi_{x,t}$ is estimated using the trapezium approximation rule and the graduated values of ρ and ν. Extrapolations are required to cater for the first week of sickness and it is necessary to ensure that the trapezium rule is based on a suitably small interval when evaluating this integral. Further details are provided by Renshaw and Haberman (1993). Adjustments are also made to allow for the growth rate in the business and to reflect the appropriate age at the start of the sickness period (leading to a claim). The values of $\pi_{x,f}$ are similarly estimated using equation (4.12).

Graduations

Following CMIR12 (1991), the graduation formula

$$\sigma_x = \exp\left(\sum_{j=0}^{s} \beta_j x^j\right) \tag{4.13}$$

is applied separately to each of the deferred periods. Investigations show that it is adequate to use quadratic predictors (i.e. $s = 2$) for each deferred period, otherwise the parameter estimates become statistically non-significant. The parameter estimates and their standard errors, for each deferred period, are presented in Table 4.12 and, for purposes of illustration, the resulting graduation for deferred period four weeks is analysed in Table 4.13. In modelling each of the deferred period data sets, the 'survival' probabilities $\pi_{x,f}$ defined by equation (4.12) and needed to form part of the offset terms have been constructed using the ρ-graduations for deferred period one week. This feature is discussed further in the appendix at the end of this chapter. Diagnostic checks, based on the deviance residual plots, and the formal tests of a graduation are of a satisfactory nature but are not reproduced here.

Table 4.12 Parameter estimates, standard errors, σ-graduations, various deferred periods

	1 week d.p.	4 week d.p.	13 week d.p.	26 week d.p.
β_0	1.7110	3.2944	5.0694	6.4587
	(0.2146)	(0.5002)	(1.0915)	(1.3735)
β_1	−0.1192	−0.2089	−0.2855	−0.3750
	(0.01025)	(0.02354)	(0.04939)	(0.06103)
β_2	0.001145	0.002137	0.002763	0.003833
	(0.0001165)	(0.0002671)	(0.0005414)	(0.0006532)

Discussion

One of the important aspects of the σ-graduations described here and in CMIR12 (1991) is the multiplicity of assumptions. It is important to assess their effects on the sensitivity of the graduations. It is instructive to classify these assumptions into two categories, which might be loosely termed (i) modelling assumptions, and (ii) data manipulation assumptions.

Thus, in the approach of CMIR12 (1991), category (i) comprises the Poisson modelling assumption, the normal approximation to the Poisson modelling assumption which induces the iterative fitting procedure in its wake, the choice of graduation formula, and the assumptions needed to compute the 'survival' probabilities $\pi_{x,f}$. The appendix to this chapter provides some comments on these assumptions. Category (ii) comprises the adjustments made in order to obtain the data (i_x, e_x): these assumptions are discussed fully by Renshaw and Haberman (1993).

It should be noted that the approach followed in this section is to graduate σ_x, the sickness inception intensity, rather than the product $\sigma_x \pi_{x,f}$, which can be regarded as a claim inception intensity. We regard σ_x as being a fundamental quantity which can then be widely used, for example in policy designs with different deferred periods. Other workers have instead concentrated on and attempted to graduate the product $\sigma_x \pi_{x,f}$ – for example, Mattson (1956) and Dillner (1969).

4.4.5 Concluding comments

Utilizing the full potential of the generalized linear modelling (GLM) framework, we have set down in this chapter a comprehensive and coherent methodology, building on the work of Forfar *et al.* (1988), for the graduation of the transition intensities of insurance related multiple state models.

Specification of the random component of the GLM, which in this context is based on independent Poisson response variables, provides the means of implementation and selection of a graduation. Specification of the systematic component (predictor) of the GLM, and the link function, is synonymous with the selection of a parameterized graduation

Table 4.13 Analysis of graduation of sickness inception intensities, σ_x, deferred period f = four weeks

x	e_x	σ_x	i_x	$\hat{E}(i_x)$	dev(x)	$\log(\pi_{x.f})$
20	87.51	0.9723	1.625	1.217	0.408	−4.2476
21	149.66	0.8613	1.069	2.117	−1.048	−4.1092
22	303.12	0.7662	10.165	4.360	5.805	−3.9753
23	614.88	0.6846	13.893	8.995	4.898	−3.8458
24	1013.61	0.6142	13.741	15.080	−1.339	−3.7205
25	1457.06	0.5534	22.367	22.047	0.320	−3.5994
26	1826.59	0.5008	32.102	28.121	3.981	−3.4822
27	2292.35	0.4552	21.658	35.922	−14.264	−3.3689
28	2748.15	0.4154	24.680	43.858	−19.178	−3.2593
29	3050.00	0.3808	44.189	49.605	−5.416	−3.1533
30	3264.66	0.3505	56.635	54.153	2.482	−3.0508
31	3467.10	0.3241	67.883	58.712	9.171	−2.9516
32	3378.61	0.3009	63.338	58.466	4.872	−2.8557
33	3350.78	0.2805	54.166	59.323	−5.157	−2.7629
34	3218.46	0.2627	53.350	58.370	−5.020	−2.6731
35	3019.86	0.2470	69.181	56.175	13.006	−2.5863
36	2972.23	0.2333	72.902	56.793	16.109	−2.5023
37	3038.03	0.2213	59.787	59.718	0.069	−2.4211
38	3135.14	0.2108	64.856	63.503	1.353	−2.3425
39	3155.07	0.2017	69.972	65.962	4.010	−2.2665
40	3158.64	0.1937	54.260	68.282	−14.022	−2.1930
41	3105.44	0.1869	82.461	69.544	12.917	−2.1219
42	3136.77	0.1811	69.137	72.914	−3.777	−2.0531
43	3092.02	0.1763	67.386	74.751	−7.365	−1.9866
44	3042.16	0.1723	84.105	76.657	7.448	−1.9222
45	3014.74	0.1691	78.966	79.343	−0.377	−1.8600
46	2950.45	0.1666	84.987	81.296	3.691	−1.7997
47	2841.34	0.1650	71.174	82.151	−10.977	−1.7414
48	2755.02	0.1640	89.531	83.774	5.757	−1.6851
49	2616.59	0.1637	85.999	83.887	2.112	−1.6306
50	2456.44	0.1642	80.426	83.244	−2.818	−1.5778
51	2273.36	0.1653	63.903	81.643	−17.740	−1.5267
52	2044.27	0.1672	71.498	77.995	−6.497	−1.4774
53	1834.21	0.1698	79.752	74.553	5.199	−1.4296
54	1676.43	0.1732	83.940	72.786	11.154	−1.3834
55	1491.82	0.1774	74.888	69.378	5.510	−1.3387
56	1280.25	0.1825	62.710	63.952	−1.242	−1.2955
57	1047.47	0.1885	56.262	56.370	−0.108	−1.2536
58	867.71	0.1956	44.986	50.446	−5.460	−1.2132
59	697.37	0.2038	39.830	43.934	−4.104	−1.1740
60	564.13	0.2133	46.991	38.627	8.364	−1.1361
61	480.77	0.2241	30.941	35.885	−4.944	−1.0995
62	400.29	0.2365	23.944	32.674	−8.730	−1.0640
63	333.48	0.2507	28.702	29.859	−1.157	−1.0297
64	201.16	0.2669	31.741	19.820	11.921	−0.9965

formula. The choice of graduation formula within this context is very versatile. In particular, the introduction in section 4.4.2 of break-point predictor terms into the graduation formula has proved to be particularly effective. The use of deviance profiles to locate both the number and optimum positions of the break-points or knots is a useful development.

The comprehensive and versatile nature of GLM methodology is demonstrated by its application to the graduation of all three transition intensities ρ, ν and σ for which data are available. Further, by utilizing the full potential of GLMs, it has been possible to avoid approximations to normality, thereby removing the need to resort to the additional complication of iterative fitting (as in Part C of CMIR12, 1991).

The more systematic approach adopted here both avoids the need to resort to *ad hoc* methods and reduces the number and complexity of the assumptions otherwise needed. Also, we are able to conduct a critical appraisal of the effects of the various assumptions on the sensitivity of the reported graduations. It will be seen that the graduations are highly sensitive to certain of the modelling assumptions and highly robust to other assumptions, thereby providing the deeper understanding needed for the assessment of such transition intensity graduations.

We accept that in presenting this approach and the graduations of ρ, ν and σ that result, we are only dealing with part of the problem of constructing graduated rates that can be directly used in practical applications. The next stage would be to consider in detail the extrapolation of our proposed formulae for values of age and sickness duration beyond the range of the data available. We have not considered this second part and it is likely that our graduation formulae may need adjustment (perhaps arbitrary) for extreme ages or sickness durations. For example, extrapolating the graduated values of ρ beyond the range of the data, we find that the graduated values increase with increasing z: this may be considered an undesirable feature requiring adjustments to be made before the graduation could be implemented practically.

In addition, adjustments would be needed to apply the standard, graduated intensities (or rates) to the specific circumstances of an individual insurer, reflecting the particular underwriting and claims management practices being adopted and the deviations from expected experience. In this context, we draw attention to the wide variation in actual experience between insurers operating in the same market: see Clark and Dullaway (1995) for comments in respect of the UK.

4.5 COMPUTATIONAL ASPECTS OF CONSTRUCTING A MULTIPLE STATE MODEL

In the previous sections of this chapter, we have presented a method for producing parametric graduations of σ_x, $\rho_{x.z}$ and $\nu_{x.z}$. For these results to be of practical use, it is important that they can be used to calculate premiums and reserves for PHI business, and to do this it is essential to be able to calculate the fundamental transition probabilities of Chapters 1-3, $_t p_x^{aa}$ etc., in an efficient manner.

In the context of the single decrement survival model, Waters and Wilkie (1987) have investigated possible methods for numerically estimating $_t p_x$ from known values of μ_x from the familiar starting point

$$_t p_x = \exp\left(-\int_0^t \mu_{x+s}\, ds\right).$$

They demonstrate that a computationally efficient method of proceeding is based on the recursion

$$_{x+h-x_0} p_{x_0} = _{x-x_0} p_{x_0} \cdot \left(\frac{1 - \frac{1}{2}h\mu_x}{1 + \frac{1}{2}h\mu_{x+h}}\right) \tag{4.14}$$

and the initial condition $_0 p_{x_0} = 1$.

This same algorithm can be used for the computation of the transition probabilities in the three-state multiple state model providing care is taken in the transformation of the set of simultaneous integro-differential equations (as in equations (1.66) and (1.67)). For further details, readers are referred to Section D of CMIR12 (1991).

4.6 REFERENCES AND SUGGESTIONS FOR FURTHER READING

CMIR12 (1991) provides a detailed exploration of the data, the graduation process and the resulting calculations needed to construct probabilities and monetary functions for practical implementation. Renshaw and Haberman (1993, 1995) provide more information on the alternative graduation framework presented here, based on generalized linear models.

For a full account of the parametric approach to graduating the force of mortality or the probability of dying arising from a survival model (and needed for the construction of a life table), the reader is referred to Forfar *et al.* (1988) or Renshaw (1991).

APPENDIX: MODELLING ASSUMPTIONS IN GRADUATION OF σ_x

In the approach advocated in section 4.4.4 for the graduation of σ_x, the basic modelling assumption is that the i_x are independent over-dispersed Poisson responses with

$$E(i_x) = m_x = \pi_{x.f} e_x \sigma_x, \quad \text{Var}(i_x) = \phi_x m_x$$

a log-link

$$\log(m_x) = \log(\pi_{x.f}) + \log(e_x) + \eta_x,$$

containing offsets and a linear predictor

$$\eta_x = \log(\sigma_x) = \sum_{j=0}^{s} \beta_j x^j,$$

a polynomial in x. The graduation formula is therefore

$$\sigma_x = \exp\left(\sum_{j=0}^{s} \beta_j x^j\right).$$

To fit the model, the β_js are estimated by optimizing the log-likelihood

$$\sum_{u} w_x[-m_x + i_x \log(m_x)] + \text{constant}$$

with weights $w_x = 1/\phi_x$. This approach fits in with the spirit of the graduation methods advocated by Forfar *et al.* (1988) and Renshaw (1992) but differs from the method of estimation adopted in CMIR12 (1991). There, the β_js are estimated by optimizing the log-likelihood

$$\sum_{u} w_x(i_x - m_x)^2 + \text{constant}$$

with weights $w_x = 1/(\phi_x e_x \pi_{x.f} \sigma_x)$, based on the standard normal approximation to the Poisson distribution. The method is equivalent to estimation by weighted least squares. This approximation has the added complication that the graduation target σ appears in the weights, requiring the introduction of an iterative fitting procedure.

We turn next to the role played by the 'survival' probabilities $\pi_{x.f}$ in the graduation of the sickness intensities σ_x. The predictor-link identity

$$\log(m_x) = \log(\pi_{x.f}) + \log(e_x) + \log(\sigma_x)$$

can be written

$$\sigma_x = \frac{m_x}{\pi_{x.f} e_x}$$

so that it follows, on replacing m_x by its estimate i_x, that the crude sickness inception intensities are given as

$$\tilde{\sigma}_x = \frac{i_x}{\pi_{x.f} e_x}.$$

Consequently the σ-graduations are highly sensitive to the values of $\pi_{x.f}$ (see Table 4.13). These in turn are determined by equation (4.12) which requires a detailed knowledge of both the ρ and ν-graduations over the complete range of ages, x, and for sickness durations, z, over the deferred period (0 to f weeks) where censoring has occurred. This latter requirement poses a problem in relation to the ρ-graduations. We also note that the ρ-graduations provide the dominant contribution to the integral expression for $\pi_{x,f}$ in equation (4.12).

The computation of the $\pi_{x,f}$ terms has been based on the numerical evaluation of the integral in expression (4.12) as described in section 4.4.4, using the ρ-graduation formula (4.8) for deferred period one week and the ν-graduation formula (4.10) for all the data combined. Alternatively, in the graduation of ρ, it would be possible to augment the data with data from the other deferred periods, but excluding the respective four-week 'run-in' periods immediately following commencement of the sickness payments. In order to evaluate the integral in expression (4.12), the ρ-graduation formula has been extrapolated back one week to time zero. It should be noted that the value of $\pi_{x,f}$ is then very sensitive to the extrapolation approach adopted, especially for the case of deferred period $f = 1$ week.

5

Critical illness cover

5.1 TYPES OF BENEFITS; ADDITIONAL BENEFITS AND ACCELERATION BENEFITS

A dread disease (DD, or 'critical illness') policy provides the policyholder with a lump sum in case of dread disease, i.e. when he is diagnosed as having an illness included in a set of diseases specified by the policy conditions. The most commonly covered diseases are heart attack, coronary artery disease requiring surgery, cancer and stroke.

The first modern DD policies appeared in South Africa in about 1983. The policy design was simple, providing a rider benefit for a temporary or endowment assurance in the event that the insured was diagnosed as having one of a set of specified conditions. Since then, critical illness policies have appeared in a number of countries including the UK and Ireland, Australia, Japan, Israel, Korea and Taiwan, and the scope of the policies has widened. Benefits are now available on an individual and group basis. Individual policies comprise two principal types: (a) **rider benefits** for a basic life policy, or (b) **stand-alone** cover. Type (a) itself takes either of two main forms: it may provide an **acceleration** of all or part of the basic life cover, or it may be an **additional** benefit.

An important point to note is that the benefit is paid on diagnosis of a specified condition, rather than on disablement. Thus, unlike other types of policy discussed, for example, in Chapter 3, the critical illness policy differs in its objectives in that it does not meet any specific need, nor does it indemnify the policyholder against any specific financial loss (for example, loss of earnings or reimbursement of medical or other expenses incurred).

5.2 A GENERAL MULTIPLE STATE MODEL; THE TIME-CONTINUOUS APPROACH

5.2.1 Probabilistic structure

Actuarial models for DD insurance can be built up starting from a

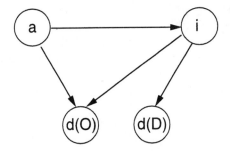

Figure 5.1 A multiple state model for critical illness covers.

multistate structure. In Fig. 5.1 the following states are considered:

a = 'active' or 'healthy';

i = 'ill', i.e. dread disease sufferer;

$d(D)$ = 'dead', death being due to dread disease;

$d(O)$ = 'dead', death being due to other causes.

Note that 'sudden deaths' due to dread disease are represented by the pair of transitions $a \to i$, $i \to d(D)$; in particular, this means that all dread disease deaths are represented by the state $d(D)$. Moreover, note that the state i is considered 'irreversible': in practice, this means that the DD cover ends after diagnosis and lump sum payment.

The following intensities (for which the notation of Chapter 3 is used) define the time-continuous probabilistic structure:

$$\mu_y^{ai} = \text{intensity of transition } a \to i;$$

$$\mu_y^{ad(O)} = \text{intensity of transition } a \to d(O);$$

$$\mu_{y,r}^{id(O)} = \text{intensity of transition } i \to d(O);$$

$$\mu_{y,r}^{id(D)} = \text{intensity of transition } i \to d(D).$$

The attained age is denoted by y; note that $\mu_{y,r}^{id(O)}$ and $\mu_{y,r}^{id(D)}$ are inception-select transition intensities, in which r denotes the time elapsed since DD inception: actually, it seems reasonable to assume that the probability of death for a dread disease sufferer depends also on the duration of the disease.

As we will see at section 5.2.4, the consideration of the states $d(O)$ and $d(D)$ (instead of the single state d) and hence of the two forces of mortality $\mu^{id(O)}$ and $\mu^{id(D)}$ can help in defining approximations and simplifying assumptions. Conversely, as far as the mortality of DD sufferers is concerned, the calculation of actuarial values only requires the use of the total force of mortality $\mu^{id(O)} + \mu^{id(D)}$, since benefits are usually independent of the cause of death.

The above definitions lead to several probabilities (see sections 1.5 and 1.6 in Chapter 1). Those of interest for actuarial calculations are:

$$_tp_y^{\overline{aa}} = \exp\left[-\int_0^t (\mu_{y+u}^{ai} + \mu_{y+u}^{ad(O)})\, du\right] \tag{5.1}$$

$$_\tau p_{y+u.0}^{\overline{ii}} = \exp\left[-\int_0^\tau (\mu_{y+u+r,r}^{id(O)} + \mu_{y+u+r,r}^{id(D)})\, dr\right]. \tag{5.2}$$

Note that $_tp_y^{\overline{aa}} = {}_tp_y^{aa}$.

5.2.2 Stand-alone cover and additional benefit

Let n denote the policy term and x the age at policy issue. Let $\bar{A}_{x,n}^{(DD)}$ denote the actuarial value at time 0 (the insured being healthy, i.e. in state a, at that time) of a lump sum benefit of 1 monetary unit at DD diagnosis, thus payable if the transition $a \rightarrow i$ occurs. Then we have:

$$\bar{A}_{x,n}^{(DD)} = \int_0^n {}_up_x^{\overline{aa}} \mu_{x+u}^{ai}\, v^u\, du. \tag{5.3}$$

Of course $\bar{A}_{x,n}^{(DD)}$ also represents the single premium according to the equivalence principle. It should be noted that this single premium may refer to a stand-alone cover as well as to an additional benefit (e.g. a rider benefit for a temporary assurance). Note that the expression on the right-hand side of (5.3) is equal to the expression used in calculating the single premium for a lump sum benefit in the case of permanent disability, when no qualification period is requested (see formula (3.87) in section 3.5).

As far as a continuous premium at constant intensity $\bar{P}_{x:n|}$ is concerned, the equivalence principle is fulfilled if:

$$\bar{P}_{x:n|} = \frac{\bar{A}_{x,n}^{(DD)}}{\bar{a}_{x:n|}^{aa}}. \tag{5.4}$$

Note that it is reasonable to assume that the premium is payable while the insured is healthy, even if a DD rider benefit is concerned such that the insurance policy does not cease at DD diagnosis.

In the case of a continuous constant premium $\bar{P}_{x:n|}$, the reserve relating to the healthy state is given by:

$$\bar{V}_{x+t,n-t}^a = \bar{A}_{x+t,n-t}^{(DD)} - \bar{P}_{x:n|}\bar{a}_{x+t:n-t|}^{aa} \tag{5.5}$$

where

$$\bar{A}_{x+t,n-t}^{(DD)} = \int_t^n {}_{u-t}p_{x+t}^{\overline{aa}} \mu_{x+u}^{ai}\, v^{u-t}\, du. \tag{5.6}$$

As regards state i, we have that:

- in the case of a stand-alone cover, the insurance policy ceases immediately after the payment of the sum assured;
- in the case of an additional benefit, only the temporary assurance remains in force (and the relevant reserve should be calculated conditional on the state i, i.e. according to the force of mortality $\mu_{y,r}^{id(O)} + \mu_{y,r}^{id(D)}$);

hence, in both cases no DD reserve relates to state i.

In the case of an additional benefit, it is important to avoid a situation of 'overpayment' (which implies higher expected costs and hence higher premiums), that could take place when death occurs within a very short period after disease inception. A solution is achieved by replacing the single cash payment with a series of payments (two or three, for example) conditional on the survival of the insured.

Let us consider a DD cover with an insured amount of 1 in total and a series of three payments at ages $x + u$ (DD diagnosis), $x + u + \tau'$ and $x + u + \tau' + \tau''$, whose proportionate amounts are $\alpha, \beta, 1 - \alpha - \beta$ respectively. The second and third payment are made if the insured is still alive. In this case the actuarial value of the benefits is given by:

$$\bar{A}_{x,n}^{(DD:\alpha,\beta)} = \int_0^n {}_u p_x^{aa} \mu_{x+u}^{ai} [\alpha v^u + {}_{\tau'} p_{x+u,0}^{ii} \beta v^{u+\tau'}$$
$$+ {}_{\tau'+\tau''} p_{x+u,0}^{ii} (1 - \alpha - \beta) v^{u+\tau'+\tau''}] \, du. \tag{5.7}$$

Note that the insured is entitled to benefits if DD diagnosis occurs within the policy term (i.e. if $u \leq n$); however, the third payment and possibly the second payment might be deferred beyond the policy term.

The active reserve, in the case of a continuous constant premium, is given by:

$$\bar{V}_{x+t,n-t}^a = \bar{A}_{x+t,n-t}^{(DD:\alpha,\beta)} - \bar{P}_{x:n\rceil} \bar{a}_{x+t:n-t\rceil}^{aa} \tag{5.8}$$

with

$$\bar{P}_{x:n\rceil} = \frac{\bar{A}_{x,n}^{(DD:\alpha,\beta)}}{\bar{a}_{x:n\rceil}^{aa}}. \tag{5.9}$$

This benefit arrangement implies reserving also in the DD state. We have:

$$\bar{V}_{x+t}^i = {}_{u+\tau'-t} p_{x+t,t-u}^{ii} \beta v^{u+\tau'-t} + {}_{u+\tau'+\tau''-t} p_{x+t,t-u}^{ii} (1 - \alpha - \beta) v^{u+\tau'+\tau''-t}$$
$$\text{if } u < t < u + \tau'; \tag{5.10'}$$

$$\bar{V}_{x+t}^i = {}_{u+\tau''-t} p_{x+t,t-u}^{ii} (1 - \alpha - \beta) v^{u+\tau''-t} \quad \text{if } u + \tau' \leq t < u + \tau' + \tau''. \tag{5.10''}$$

A different solution consists in paying the whole DD benefit after a fixed qualification period (e.g. one or two months). Let τ denote the length of the qualification period. In this case, the actuarial value is simply given by:

$$\bar{A}^{(DD)}_{x,n(\tau)} = \int_0^n {}_up_x^{aa}\mu^{ai}_{x+u} \cdot {}_\tau p^{ii}_{x+u,0}v^{u+\tau}\,du. \tag{5.11}$$

Formula (5.11) represents a particular case of (5.7), namely with

$$\alpha = 0, \quad \beta = 1, \quad \tau' = \tau.$$

Note that the expression on the right-hand side of (5.11) is equal to the expression used in calculating the single premium for a lump sum benefit in case of permanent disability, with a qualification period of τ years (see formula (3.186) in section 3.5).

This arrangement implies reserving the DD state, namely during the qualification period. The expression for the reserve can be easily derived from equation (5.10'), letting $\alpha = 0$, $\beta = 1$, $\tau' = \tau$ (of course, τ'' is meaningless).

5.2.3 Acceleration benefit

Let us consider a temporary assurance with a DD acceleration benefit. Let us assume that, for a given sum assured S, the amount λS ($0 < \lambda \leq 1$) is payable on DD diagnosis, while the remaining amount $(1 - \lambda)S$ is payable on death, if this occurs within the policy term n. Thus λ represents the 'acceleration parameter'.

Figure 5.2, in which it is assumed $S = 1$, can help in understanding the policy design and in interpreting the following formulae. Note that if $\lambda = 1$, then the graph 'loses' the arcs $i \rightarrow d(O)$ and $i \rightarrow d(D)$; in this case state i is an absorbing state, since the whole insurance cover ceases after transition $a \rightarrow i$.

For brevity we now restrict our attention to premium calculation only. Let us assume $S = 1$. The actuarial value of the benefits, i.e. relating to the

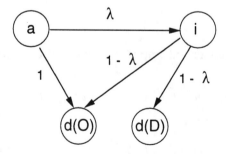

Figure 5.2 A multiple state model for a temporary assurance with a DD acceleration benefit.

death benefit (D) and to the DD acceleration benefit as well, is given by:

$$\bar{A}_{x,n}^{(D+DD:\lambda)} = \int_0^n {}_u p_x^{\overline{aa}} \left[\mu_{x+u}^{ad(O)} v^u + \mu_{x+u}^{ai} \left(\lambda v^u + (1-\lambda) \right. \right.$$

$$\left. \left. \times \int_0^{n-u} {}_r p_{x+u,0}^{\overline{ii}} (\mu_{x+u+r,r}^{id(O)} + \mu_{x+u+r,r}^{id(D)}) v^{u+r} \, dr \right) \right] du. \quad (5.12)$$

In particular, with $\lambda = 1$ we obtain:

$$\bar{A}_{x,n}^{(D+DD:1)} = \int_0^n {}_u p_x^{\overline{aa}} [\mu_{x+u}^{ad(O)} + \mu_{x+u}^{ai}] v^u \, du. \quad (5.13)$$

Let $\bar{P}_{x:n\rceil}$ denote the continuous constant premium meeting the death benefit as well as the acceleration benefit. Assuming that premiums are payable while the insured is healthy, the equivalence principle is fulfilled if:

$$\bar{P}_{x:n\rceil} = \frac{\bar{A}_{x,n}^{(D+DD:\lambda)}}{\bar{a}_{x:n\rceil}^{\overline{aa}}}. \quad (5.14)$$

A decomposition of the actuarial value given by equation (5.12) is straightforward. Let us consider the following quantities:

$$\bar{A}_{x,n}^{(D:\lambda)} = \int_0^n {}_u p_x^{\overline{aa}} \left[\mu_{x+u}^{ad(O)} v^u + \mu_{x+u}^{ai} (1-\lambda) \right.$$

$$\left. \times \int_0^{n-u} {}_r p_{x+u,0}^{\overline{ii}} (\mu_{x+u+r,r}^{id(O)} + \mu_{x+u+r,r}^{id(D)}) v^{u+r} \, dr \right] du \quad (5.15)$$

$$\bar{A}_{x,n}^{(DD:\lambda)} = \int_0^n {}_u p_x^{\overline{aa}} \mu_{x+u}^{ai} \lambda v^u \, du. \quad (5.16)$$

Then:

$$\bar{A}_{x,n}^{(D+DD:\lambda)} = \bar{A}_{x,n}^{(D:\lambda)} + \bar{A}_{x,n}^{(DD:\lambda)}. \quad (5.17)$$

Note that $\bar{A}_{x,n}^{(DD:\lambda)}$ is the actuarial value of the acceleration benefit. Conversely, $\bar{A}_{x,n}^{(D:\lambda)}$ represents the actuarial value of the death benefit. In this actuarial value the mortality of both healthy lives (with benefit equal to 1) and DD sufferers (with benefit equal to $1-\lambda$) is considered. With $\lambda = 1$, equations (5.15) and (5.16) respectively become:

$$\bar{A}_{x,n}^{(D:1)} = \int_0^n {}_u p_x^{\overline{aa}} \mu_{x+u}^{ad(O)} v^u \, du \quad (5.18)$$

$$\bar{A}_{x,n}^{(DD:1)} = \int_0^n {}_u p_x^{\overline{aa}} \mu_{x+u}^{ai} v^u \, du. \quad (5.19)$$

Note that $\bar{A}_{x,n}^{(D:1)}$ represents the single premium of a temporary assurance in which the death benefit is restricted to insureds who are not DD sufferers. Conversely, $\bar{A}_{x.n}^{(DD;1)}$ coincides with the single premium, $\bar{A}_{x.n}^{(DD)}$, of a stand-alone benefit or an additional benefit (see equation (5.3)).

Remark
It should be pointed out that formulae (5.7), (5.11) and (5.12), in particular, witness the flexibility of the time-continuous multiple state approach in developing formulae for evaluating various benefit arrangements. However, it should also be stressed that the estimation of the relevant probabilities and intensities may be a difficult matter owing to the current paucity of data (some simplifying hypotheses are dealt with in section 5.2.4).

5.2.4 Approximations for practical implementations

The amount of insured experience (in each of the markets where these products are sold) is limited and this means that the practical implementation of formulae like (5.3), (5.7), (5.11) or (5.12) will need both approximations and the use of routinely collected population-based data (for example, on the incidence of a particular critical illness in the population and on the mortality rates of those suffering from such a critical illness). The details will depend on the statistics available on the national population being considered.

A common set of assumptions used is that

$$\mu_{y,r}^{id(O)} = \mu_y^{id(O)} \qquad (5.20)$$

$$\mu_{y,r}^{id(D)} = \mu_y^{id(D)} \qquad (5.21)$$

i.e. that the mortality transition intensities from state i are independent of the duration of the disease, r. It is also common to consider the ratio of the mortality transition intensities for causes of death other than the critical disease for healthy and ill lives, i.e. use the factor $1 + m_y$ where

$$\mu_y^{id(O)} = \mu_y^{ad(O)}(1 + m_y). \qquad (5.22)$$

The approach widely adopted by practitioners is to use disease specific incidence data for the computation of μ_y^{ai}, and to introduce data on the proportion of all deaths at age y in the population that are attributable to the critical disease, k_y, and require the following approximation to hold:

$$\mu_y^{ad(O)} + \mu_y^{id(O)} = (1 - k_y)\mu_y \qquad (5.23)$$

where μ_y is the total force of mortality in the population at age y, allowing for all causes of death (and ignoring the structure of the multiple state model).

The simplifying assumption that $m_y = 0$ can also be introduced on the grounds that the resulting premiums are relatively insensitive to the choice of m_y providing that age y is not too advanced: it must be pointed out that the choice $m_y = 0$ tends to simplify the calculations and to incorporate a margin in that the resulting premium rates are higher than for the case $m_y \neq 0$. Reliable data on $\mu^{id(O)}$ and $\mu^{id(D)}$ may not be directly available: much depends on the quality of the data emerging from cancer registries, *inter alia*.

Care needs to be taken in using population-based data for estimating the parameters in the multiple state model which is intended for applicability to insured lives. Adjustments (possibly *ad hoc*) will be needed to reflect the selection processes operating among a portfolio of insured lives. The combination of incidence rates for different critical conditions, for determining μ_y^{ai}, also needs consideration: the simple addition of transition intensities implies that the probability of an individual experiencing two (or more) unrelated insured events is low. This assumption might be reasonable in pricing but will lead to an overstatement in the expected value of benefits, particularly if the underlying policy continues to advanced ages (as is common in the UK, for example).

5.3 THE TIME-DISCRETE APPROACH

In this section we present a time-discrete Markov model for critical illness covers. States and transitions are still represented by the graph depicted in Fig. 5.1. A time-discrete probabilistic structure for disability benefits has been already described in Chapter 2, sections 2.1 to 2.3, and in Chapter 3, section 3.2.7. So we can now restrict our attention to some aspects only concerning DD covers.

5.3.1 Probabilities for time-discrete models

Let x denote the (integer) age at policy entry. We now consider the process $\{S(y); y = x, x+1, \ldots\}$ assuming that it is a time-discrete inhomogeneous Markov chain.

Let us consider the following probabilities related to a healthy insured aged y.

w_y = probability of becoming a DD sufferer within one year;

p_y^{aa} = probability of being healthy at age $y + 1$;

q_y^{aa} = probability of dying within one year, the death occurring in state a;

p_y^{ai} = probability of being a DD sufferer at age $y + 1$;

$q_y^{ai(D)}$ = probability of becoming a DD sufferer and dying within one year, due to DD;

$q_y^{ai(O)}$ = probability of becoming a DD sufferer and dying within one year, due to other causes;

q_y^{ai} = probability of becoming a DD sufferer and dying within one year;

p_y^a = probability of being alive at age $y + 1$;

q_y^a = probability of dying within one year.

Let us consider the following probabilities related to a DD sufferer aged y. Remember that the state i is considered 'irreversible' and hence transitions $i \rightarrow a$ are not admitted (see Fig. 5.1):

p_y^i = probability of being alive at age $y + 1$;

$q_y^{i(D)}$ = probability of dying within one year, due to DD;

$q_y^{i(O)}$ = probability of dying within one year, due to other causes;

q_y^i = probability of dying within one year.

It is interesting to express the events considered in defining the above probabilities in terms of transitions or sequences of transitions occurring during the year (see Fig. 5.1). For example:

$$q_y^{aa} \text{ refers to } a \rightarrow d(O);$$

$$p_y^{ai} \text{ refers to } a \rightarrow i;$$

$$q_y^{ai(D)} \text{ refers to } a \rightarrow i, i \rightarrow d(D);$$

$$q_y^{ai(O)} \text{ refers to } a \rightarrow i, i \rightarrow d(O);$$

$$q_y^{i(D)} \text{ refers to } i \rightarrow d(D).$$

Several relations hold among the probabilities defined above; for example:

$$q_y^{ai} = q_y^{ai(O)} + q_y^{ai(D)}; \tag{5.24}$$

$$w_y = p_y^{ai} + q_y^{ai}; \tag{5.25}$$

$$q_y^a = q_y^{aa} + q_y^{ai}; \tag{5.26}$$

$$w_y + p_y^{aa} + q_y^{aa} = 1; \tag{5.27}$$

$$q_y^i = q_y^{i(O)} + q_y^{i(D)}; \tag{5.28}$$

$$p_y^i + q_y^i = 1. \tag{5.29}$$

Premium calculation, according to formulae that will be presented in sections 5.3.2 and 5.3.3, only requires a subset of the probabilities defined above. Nevertheless, the definition of the whole probabilistic structure (consistent with the graph depicted in Fig. 5.1) can help the reader in understanding approximations and simplifying assumptions in a time-discrete context, like those proposed by Dash and Grimshaw (1990).

5.3.2 Premiums and reserve for the additional benefit

Let us consider an additional benefit, for example a rider to a temporary assurance. Let $_hp_x^{\overline{aa}}$ denote the probability of remaining healthy for h years (at least). Since state i is irreversible, we simply have:

$$_hp_x^{\overline{aa}} = p_x^{aa} p_{x+1}^{aa} \cdots p_{x+h-1}^{aa} \tag{5.30}$$

with $_0p_x^{\overline{aa}} = 1$ (note that we also have $_hp_x^{\overline{aa}} = {}_hp_x^{aa}$).

Assuming a uniform distribution of the DD onset over each policy year and assuming that the sum assured (equal to 1 monetary unit) is payable at DD onset, the actuarial value of the DD benefit is given by:

$$A_{x,n}^{(DD)} = \sum_{h=0}^{n-1} {}_hp_x^{\overline{aa}} w_{x+h} v^{h+1/2}. \tag{5.31}$$

Of course, the single premium fulfilling the equivalence principle is given by $A_{x,n}^{(DD)}$. The annual level premium, $P_{x:n\rceil}$, payable while the insured is healthy, is given by:

$$P_{x:n\rceil} = \frac{A_{x,n}^{(DD)}}{\ddot{a}_{x:n\rceil}^{aa}} \tag{5.32}$$

where

$$\ddot{a}_{x:n\rceil}^{aa} = \sum_{h=0}^{n-1} {}_hp_x^{\overline{aa}} v^h.$$

Note that the single and the annual premium of an additional benefit respectively represent also the single and the annual premium of a stand-alone DD cover.

In the case of annual constant premiums, the reserve relating to the healthy state is given by:

$$V_{x+t,n-t}^a = A_{x+t,n-t}^{(DD)} - P_{x:n\rceil} \ddot{a}_{x+t:n-t\rceil}^{aa} \tag{5.33}$$

with a straightforward definition for $A_{x+t,n-t}^{(DD)}$. No DD reserve relates to state i; to this purpose, the reader is referred to section 5.2.2.

5.3.3 Premiums for the acceleration benefit

For simplicity we shall restrict our attention to an acceleration benefit with $\lambda = 1$. Note that the general case $(0 < \lambda \leq 1)$ has been dealt with, in a time-continuous framework, in section 5.2.3. Moreover, for the sake of brevity we shall restrict our discussion to premium calculation only.

First, let us calculate the single premium meeting the death benefit as well as the DD acceleration benefit. With the usual assumptions (and consistent with the notation used in section 5.2), we have:

$$A_{x,n}^{(D+DD:1)} = \sum_{h=0}^{n-1} {}_h p_x^{aa} (q_{x+h}^{aa} + w_{x+h}) v^{h+1/2}. \qquad (5.34)$$

The relevant annual premium is obviously given by:

$$P_{x:n\rceil} = \frac{A_{x,n}^{(D+DD:1)}}{\ddot{a}_{x:n\rceil}^{aa}}. \qquad (5.35)$$

Some decompositions of the single premium expressed by equation (5.34) are of interest. For simplicity, let us start with the case $n = 1$. From (5.34) we immediately obtain:

$$A_{x,1}^{(D+DD:1)} = (q_x^{aa} + w_x) v^{1/2} = q_x^{aa} v^{1/2} + w_x v^{1/2}. \qquad (5.36)$$

The first term, $q_x^{aa} v^{1/2}$, is the single premium of a temporary assurance, in which the death benefit is restricted to non-DD sufferers. The second term is clearly the single premium of the DD benefit.

An alternative decomposition can be found using equations (5.25) and (5.26). We obtain:

$$A_{x,1}^{(D+DD:1)} = q_x^{aa} v^{1/2} + w_x v^{1/2} = q_x^a v^{1/2} + (w_x + q_x^{aa} - q_x^a) v^{1/2}$$

$$= q_x^a v^{1/2} + (w_x - q_x^{ai}) v^{1/2} = q_x^a v^{1/2} + p_x^{ai} v^{1/2}. \qquad (5.37)$$

Now the first term, $q_x^a v^{1/2}$, represents the single premium of a conventional one-year assurance, in which the mortality has been evaluated taking into account that the insured is in state a at policy issue. The second term is the actuarial value of the DD acceleration benefit, excluding the actuarial value of the benefit itself in the case of death within the year (this value being included in the first term, since $q_y^a = q_y^{aa} + q_y^{ai}$).

Let us turn to the general case $(n \geq 1)$. From equation (5.34) we get:

$$A_{x,n}^{(D+DD:1)} = \sum_{h=0}^{n-1} {}_h p_x^{aa} q_{x+h}^{aa} v^{h+1/2} + \sum_{h=0}^{n-1} {}_h p_x^{aa} w_{x+h} v^{h+1/2}. \qquad (5.38)$$

Note that equation (5.38) is the time-discrete counterpart of the decomposition expressed, in a time-continuous context, by equations (5.18) and (5.19). As in the case $n = 1$ (see equation (5.36)), the first term is the single premium of a temporary assurance, in which the death benefit is restricted to non-DD sufferers. The second term is clearly the single premium of the DD benefit.

An alternative decomposition is as follows:

$$A_{x,n}^{(D+DD:1)} = \sum_{h=0}^{n-1} {}_h p_{\overline{x}}^{aa}(q_{x+h}^a - q_{x+h}^{ai} + p_{x+h}^{ai} + q_{x+h}^{ai})v^{h+1/2}$$

$$= \sum_{h=0}^{n-1} {}_h p_{\overline{x}}^{aa} q_{x+h}^a v^{h+1/2} + \sum_{h=0}^{n-1} {}_h p_{\overline{x}}^{aa} p_{x+h}^{ai} v^{h+1/2}. \qquad (5.39)$$

Decomposition (5.39) can be easily interpreted, representing a generalization of (5.37).

For brevity, we omit formulae for annual premiums; their decomposition is then straightforward in the light of equations (5.38) and (5.39).

5.4 REFERENCES AND SUGGESTIONS FOR FURTHER READING

Critical illness insurance, in the form of stand-alone cover as well as in the form of rider benefit for temporary and whole life policies, provides an important type of benefit in the field of insurances of the person. Nevertheless, the relevant actuarial literature is not very extensive.

Critical illness products and related benefit design aspects are presented in the handbooks by Frankona Rückversicherung (1989), Mercantile & General Reinsurance (1992) and Münchener Rück (1989).

Actuarial aspects are discussed by Allerdissen *et al.* (1993), Dash and Grimshaw (1990) and Fabrizio and Gratton (1994). In particular, the paper by Dash and Grimshaw (1990) discusses approximations and simplifying assumptions which can be adopted, within a time-discrete context, in pricing DD products (see section 5.2.4). Finally, the paper by Lörper *et al.* (1991) is particularly devoted to problems concerning the statistical bases.

6

Long-term care insurance

6.1 TYPES OF BENEFITS

Long-term care (LTC) is care that is required in relation to chronic (or long-lasting) conditions or ailments. LTC insurance provides income support for the insured, who needs nursing and/or medical care. As far as actuarial calculations are concerned, LTC benefits can be classified into three categories:

1. fixed amount annuities (possibly depending on the frailty level) sold to healthy people;
2. fixed amount annuities (possibly depending on the frailty level) sold to elderly people at the time of entering into a nursing home or to current residents of nursing care homes;
3. nursing and medical expense refunding.

A fourth type of LTC cover can be added, according to which the insured can choose, in the case of a claim, between an annuity benefit or an appropriate care service provided by organizations offering nursing care services. This product is sold in Japan, and is linked to either a health insurance or pension plan.

Multiple state modelling provides a flexible tool for expressing the actuarial structure of LTC products. In particular, benefits of type 1 and 2 are considered in this chapter. Some examples of products belonging to these categories are as follows.

1. **Stand-alone policy**, providing a fixed amount annuity in the case of LTC need. The amount of the annuity can be defined as a function of the frailty level.
2. **Enhanced annuity**. An annuity is sold to an elderly person at the time of entering a nursing home, or to current residents of nursing care homes. The enhancement in the annuity payment, compared with a standard life annuity, comes from the use of higher mortality assumptions because of the person's health status.

3. **LTC cover as a rider benefit**. Usually the rider is considered as an accelerated death benefit or 'prepayment' benefit. A common method consists in adding to a whole life cover (for example a traditional universal life product) a monthly benefit of, say, 2% of the sum assured, payable for 50 months at most. An additional feature would be to add a guarantee that the benefit would continue to be paid when an LTC claim exceeded 50 months.

4. **Enhanced pension**. This product is sold at retirement date and is a combination of a standard pension annuity paid while the policyholder is healthy, and an uplifted income paid while the policyholder is claiming LTC benefits. For a given amount of single premium, the 'price' of the uplift is a reduction in the initial pension income.

5. **Insurance packages**, usually comprising the following benefits:

    - LTC annuity (whose amount can be defined as a function of the frailty level);
    - deferred life annuity (e.g. from age 80), while the insured is not claiming LTC benefits;
    - lump sum benefit on death.

    An appealing insurance package, so-called **whole life health insurance**, consists of a PHI cover providing pre-retirement benefits in the case of disability on an 'own or similar occupation' basis and an LTC annuity whose amount possibly depends on the frailty level. For example, packaging LTC covers with PHI is common in Israel.

In order to avoid (or limit) devaluation of the insurance cover by inflation, some insurers offer index-linked schemes for benefits and premiums. The problem of indexing benefits in the field of insurances of the person will be dealt with in Chapter 8, in the general framework of multiple state modelling.

From the viewpoint of policy design, a definition of frailty is needed in order to determine whether or not an insured is eligible to claim LTC benefits. Such a definition is often called 'benefit trigger'. The frailty and hence the need for LTC are a consequence of an individual's ability to take care of himself by performing the basic tasks of everyday independent living. Then, the extent of that frailty can be measured by the **ADL** (activities of daily living) criteria. Some ADLs which are commonly considered are: eating, bathing, moving around, going to the toilet, dressing.

The number of ADLs used, their exact definition, and the minimum number which must be failed to entitle a policyholder to benefits, vary by company and country. Using an ADL system also allows the insurer to offer scaled benefits (i.e. benefits whose amounts depend on the frailty level; see for example points 1 and 5), by requiring more ADLs to be failed for the higher levels of benefits.

LTC covers were first developed in the US in the mid-1970s. In the early days of LTC insurance there, various gatekeeper mechanisms were used, such as the requirement that a certain number of days in the hospital witness medical necessity. These requirements have mostly been eliminated.

In 1993 LTC insurance was offered by 118 US insurance companies, and sales of policies have grown by 27% p.a. compound between 1987 and 1993. In Germany, LTC products became available in 1985. Today these products are also available in other industrialized countries, e.g. the UK, France, Switzerland, the Netherlands, Japan, Israel, South Africa.

Two of the critical factors contributing to the success and growth of LTC insurance are, as noted by Gatenby (1991), the relative level of state provision and the tax treatment of LTC contracts, in terms of deductibility of the premiums and the status of the benefit payments. As an example, recent legislative changes in the US mean that LTC premiums there may be treated as tax deductible (as for accident and health insurance), depending on whether the contract meets certain conditions. Hauser and Litow (1996) anticipate a further boost to sales from these changes and a possible increase in the number of companies operating in the market.

A further type of LTC provision occurs within continuing care retirement communities (CCRCs) which have become established in the US. CCRCs offer housing and a range of other services, including long-term care. The cost is usually met by a combination of entrance charge plus periodic fees (i.e. single premium plus monthly premiums). As noted by Jones (1997), a simple multiple state model for the level of care provided to a resident of a CCRC would comprise the four states: alive and independent, occupying a skilled nursing facility (SNF) on a 'temporary' basis, occupying an SNF on a 'permanent' basis, and dead. We will not consider CCRCs further in this book. Readers are referred to Winklevoss and Powell (1984) for a discussion of practical issues associated with the management of CCRCs and to Jones (1995, 1996, 1997a) for a discussion of important modelling issues; statistical aspects are dealt with by Jones (1997b).

6.2 A GENERAL MULTIPLE STATE MODEL FOR LTC BENEFITS –
    TIME-CONTINUOUS APPROACH

Since several types of benefits will be dealt with (LTC annuities, standard annuities, death benefits), in this chapter, unlike Chapters 3 and 5, we will use the general notation defined in Chapter 1.

Let us consider a multiple state model which consists of the following states:

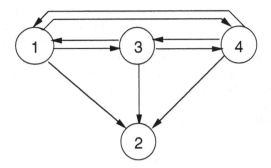

Figure 6.1 A general multiple state model for LTC benefits.

state 1 = 'healthy';

state 2 = 'dead';

state 3 = 'frailty at level I';

state 4 = 'frailty at level II';

$\vdots$

state N = 'frailty at level $N - 2$'.

As regards the number of frailty (or disability) or LTC states, we will restrict our attention to models with two LTC states at most. The model depicted in Fig. 6.1 concerns the case of two LTC states. As we will see in the next section, any multiple state model representing a particular LTC product (a stand-alone cover as well as more complicated policies) is a particular sub-graph of the graph depicted in Fig. 6.1. It is understood that when just one LTC state is considered, this state includes all frailty levels entitling the insured to benefits.

According to the transition intensity approach, nine intensities should in principle be assigned in order to define the probabilistic structure of this model. In practice, it is important to stress that:

- deriving transition probabilities from the relevant set of simultaneous differential equations is a very hard computational problem;
- the implementation of such a general multistate model requires a considerable amount of statistical data, whereas available LTC data are rather scanty.

Hence, it is necessary to resort to simpler multiple state models. In particular, because of the usually chronic character of frailty, it is reasonable to disregard the possibility of recovery, at least as far as the highest levels of disability are concerned (state 4, in the example provided by Fig. 6.1).

As regards benefits, we assume that a continuous annuity benefit is paid at an instantaneous rate $b_i(t)$ while the insured is in the LTC state i. For the sake of simplicity, constant rates b_i will be assumed. Further benefits will be included according to the policy design. We assume that a continuous premium is paid at an instantaneous rate $p_1(t)$ while the insured is in the healthy state (when this assumption is consistent with the policy design). We denote by m the premium term. For simplicity we will consider a level premium $p_1(t) = p$. In some cases, the policy structure implies a single premium at time 0.

6.3 MULTIPLE STATE MODELS FOR SOME LTC COVERS

In this section we describe, in the framework of time-continuous multiple state modelling, the actuarial structure of some LTC products drawn from various insurance markets. Formulae for premium and reserves will be derived according to the equivalence principle. In section 6.4 some time-discrete formulae for premiums and reserves will be proposed.

As regards the probabilistic structure, we assume the Markov property, without splitting the LTC states, thus using non-select transition intensities.

6.3.1 Stand-alone annuity

We assume the four-state model depicted in Fig. 6.2. Hence, recoveries are disregarded. Let us consider an LTC stand-alone cover which pays a continuous annuity benefit at an instantaneous constant rate b_i while the insured is in the disability state i, $i = 3, 4$. It is reasonable to assume $b_3 < b_4$.

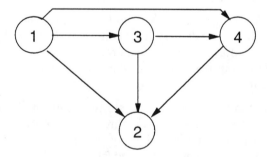

Figure 6.2 A multiple state model for LTC annuities, with two disability states.

The probabilistic structure is completely determined by assigning the following transition intensities:

$$\mu_{13}(t), \mu_{14}(t), \mu_{34}(t): \text{ intensities of 'disability'};$$

$$\mu_{12}(t), \mu_{32}(t), \mu_{42}(t): \text{ intensities of mortality}.$$

As the multiple state model only contains strictly transient states and an absorbing state, transition probabilities can be easily derived from the transition intensities by solving the relevant set of differential equations. Let $\mathscr{B}_1^{LTC}(0, \infty)$ denote the actuarial value at time 0, with $S(0) = 1$, of the LTC annuity benefits (the reference to the amounts b_3 and b_4 is understood). Then we have:

$$\mathscr{B}_1^{LTC}(0, \infty) = b_3 \int_0^{+\infty} v^z P_{13}(0, z) \, dz + b_4 \int_0^{+\infty} v^z P_{14}(0, z) \, dz \qquad (6.1)$$

i.e.

$$\mathscr{B}_1^{LTC}(0, \infty) = b_3 \bar{a}_{13}(0, \infty) + b_4 \bar{a}_{14}(0, \infty). \qquad (6.1')$$

Of course $\mathscr{B}_1^{LTC}(0, \infty)$ also represents the single premium according to the equivalence principle. As far as a continuous level premium $p_1(t) = p$ is concerned, the equivalence principle is fulfilled if:

$$p \bar{a}_{11}(0, m) = b_3 \bar{a}_{13}(0, \infty) + b_4 \bar{a}_{14}(0, \infty). \qquad (6.2)$$

Formulae (6.1) and (6.2) are based on the probabilities of being in the two LTC states (and on the probability of being healthy). A generalization of the procedure described in section 1.12.3 leads to expressing the actuarial value of benefits in terms of the probabilities of becoming and remaining disabled. To this purpose, the following relations must be considered (see for analogy equation (1.40)):

$$P_{13}(t, s) = \int_t^s P_{11}(t, u) \mu_{13}(u) P_{\underline{33}}(u, s) \, du; \qquad (6.3a)$$

$$P_{14}(t, s) = \int_t^s P_{11}(t, u) \mu_{14}(u) P_{\underline{44}}(u, s) \, du$$

$$+ \int_t^s P_{13}(t, u) \mu_{34}(u) P_{\underline{44}}(u, s) \, du. \qquad (6.3b)$$

It should be stressed that, since states 1, 3 and 4 are strictly transient states, we have

$$P_{ii}(t, s) = P_{\underline{ii}}(t, s), \quad i = 1, 3, 4.$$

Thus, in the following we will use equivalently the probabilities $P_{ii}(t, s)$ or $P_{\underline{ii}}(t, s)$.

Substituting equations (6.3a) and (6.3b) into (6.1) we find:

$$\mathscr{B}_1^{\text{LTC}}(0,\infty) = b_3 \int_0^{+\infty} P_{\underline{11}}(0,u)\mu_{13}(u)v^u \left[\int_u^{+\infty} P_{\underline{33}}(u,z)v^{z-u}\,dz \right] du$$

$$+ b_4 \int_0^{+\infty} P_{\underline{11}}(0,u)\mu_{14}(u)v^u \left[\int_u^{+\infty} P_{\underline{44}}(u,z)v^{z-u}\,dz \right] du$$

$$+ b_4 \int_0^{+\infty} P_{\underline{13}}(0,u)\mu_{34}(u)v^u \left[\int_u^{+\infty} P_{\underline{44}}(u,z)v^{z-u}\,dz \right] du$$

$$= b_3 \int_0^{+\infty} P_{\underline{11}}(0,u)\mu_{13}(u)v^u \bar{a}_{\underline{33}}(u,\infty)\,du$$

$$+ b_4 \int_0^{+\infty} P_{\underline{11}}(0,u)\mu_{14}(u)v^u \bar{a}_{\underline{44}}(u,\infty)\,du$$

$$+ b_4 \int_0^{+\infty} P_{\underline{13}}(0,u)\mu_{34}(u)v^u \bar{a}_{\underline{44}}(u,\infty)\,du \qquad (6.4)$$

where $\bar{a}_{ii}(u,\infty)$ denotes the actuarial value of a lifetime annuity paid while the insured is in LTC state i.

Note that when strictly transient states are concerned $\bar{a}_{\underline{ii}}(t,u) = \bar{a}_{ii}(t,u)$. The occupancy probabilities can be immediately derived from the set of transition intensities. We have:

$$P_{\underline{11}}(t,s) = \exp\left[-\int_t^s [\mu_{12}(z) + \mu_{13}(z) + \mu_{14}(z)]\,dz \right] \qquad (6.5)$$

$$P_{\underline{33}}(t,s) = \exp\left[-\int_t^s [\mu_{32}(z) + \mu_{34}(z)]\,dz \right] \qquad (6.6)$$

$$P_{\underline{44}}(t,s) = \exp\left[-\int_t^s \mu_{42}(z)\,dz \right]. \qquad (6.7)$$

The definitions of the actuarial values of LTC annuity benefits can be easily generalized to any time t, $t \geq 0$, and relevant states $S(t)$. We find:

$$\mathscr{B}_1^{\text{LTC}}(t,\infty) = b_3 \int_t^{+\infty} v^{z-t} P_{13}(t,z)\,dz + b_4 \int_t^{+\infty} v^{z-t} P_{14}(t,z)\,dz$$

$$= b_3 \int_t^{+\infty} P_{\underline{11}}(t,u)\mu_{13}(u)v^{u-t} \bar{a}_{\underline{33}}(u,\infty)\,du$$

$$+ b_4 \int_t^{+\infty} P_{\underline{11}}(t,u)\mu_{14}(u)v^{u-t} \bar{a}_{\underline{44}}(u,\infty)\,du$$

$$+ b_4 \int_t^{+\infty} P_{\underline{13}}(t,u)\mu_{34}(u)v^{u-t} \bar{a}_{\underline{44}}(u,\infty)\,du; \qquad (6.8)$$

$$\mathscr{B}_3^{\text{LTC}}(t,\infty) = b_3 \int_t^{+\infty} v^{z-t} P_{\underline{33}}(t,z)\, dz + b_4 \int_t^{+\infty} v^{z-t} P_{34}(t,z)\, dz$$

$$= b_3 \bar{a}_{\underline{33}}(t,\infty) + b_4 \int_t^{+\infty} P_{\underline{33}}(t,u)\mu_{34}(u)v^{u-t}\bar{a}_{\underline{44}}(u,\infty)\, du; \quad (6.9)$$

$$\mathscr{B}_4^{\text{LTC}}(t,\infty) = b_4 \int_t^{+\infty} v^{z-t} P_{\underline{44}}(t,z)\, dz = b_4 \bar{a}_{\underline{44}}(t,\infty). \quad (6.10)$$

Three reserves must be defined, relating to the three (strictly) transient states. The 'healthy' reserve (thus conditional on $S(t) = 1$) is given by:

$$\bar{V}_1(t) = \mathscr{B}_1^{\text{LTC}}(t,\infty) - p\bar{a}_{\underline{11}}(t,m) \qquad \text{if } t < m \quad (6.11)$$

$$\bar{V}_1(t) = \mathscr{B}_1^{\text{LTC}}(t,\infty) \qquad \text{if } t \geq m \quad (6.12)$$

whereas the reserves related to the two LTC states (thus conditional on $S(t) = 3$ and $S(t) = 4$, respectively) are simply given by:

$$\bar{V}_3(t) = \mathscr{B}_3^{\text{LTC}}(t,\infty) \quad (6.13)$$

$$\bar{V}_4(t) = \mathscr{B}_4^{\text{LTC}}(t,\infty). \quad (6.14)$$

6.3.2 Stand-alone annuity – a simplified model

Let us consider an LTC cover in which just one state of disability, $i = 3$, is assumed. Recoveries are in principle possible (but actually disregarded, as we will see later). See Fig. 6.3.

A time-continuous annuity benefit at an instantaneous constant rate b is payable while the insured is in state 3. At policy issue, the random present value of this annuity (see formula (1.101) in Chapter 1) is:

$$Y_0(0,\infty) = b \int_0^{+\infty} v^t I_{\{S(t)=i\}}\, dt. \quad (6.15)$$

Let M denote the random number of LTC claims. Then, the random present value $Y_0(0,\infty)$, which we now denote for convenience by Y_x,

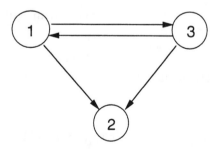

Figure 6.3 A multiple state model for LTC annuities, with one disability state.

where x is the age at entry, can also be expressed as follows:

$$Y_x = \begin{cases} b \sum_{h=1}^{M} v_{T_h}^{T_h} / \bar{a}_{D_h]} & \text{if } M \geq 1 \\ 0 & \text{if } M = 0 \end{cases} \tag{6.16}$$

where:

T_1 = random time at the beginning of the first LTC spell;

D_1 = random duration of the first LTC spell;

T_2 = random time at the beginning of the second LTC spell.

. . .

As said above, this model in principle allows for recovery (note that the model described in section 6.3.1 does not allow for this possibility). A simpler model has been proposed by Beekman (1990). For brevity we restrict our attention to single premium calculation only. Let us consider the following random variable:

$$N_x = \begin{cases} \sum_{h=1}^{M} D_h & \text{if } M \geq 1 \\ 0 & \text{if } M = 0. \end{cases} \tag{6.17}$$

The random variable N_x is the random time totally spent in the LTC state by a person aged x and healthy at policy issue. Now, let us define the following random present value:

$$Z_x = b_{T_1} / \bar{a}_{N_x]} = b v^{T_1} \bar{a}_{N_x]}; \tag{6.18}$$

of course, we have:

$$Z_x \geq Y_x$$

and then

$$E(Z_x) \geq E(Y_x) \tag{6.19}$$

because Z_x reflects the annuity payments in a regular stream, and hence at earlier times than the interrupted stream of payments in Y_x (see Fig. 6.4, in which $M = 3$).

The actuarial value (at policy issue, hence with $S(0) = 1$) of the LTC annuity benefits is given by

$$\mathscr{B}_1^{\text{LTC}}(0, \infty) = E(Y_x) \tag{6.20}$$

so that it can be approximated by

$$\mathscr{B}_1^{\text{LTC}}(0, \infty) \cong E(Z_x)$$

$$= b \int_0^{+\infty} P_{\underline{11}}(0, u) \mu_{13}(u) v^u E[\bar{a}_{N_x]} | (S(0) = 1) \wedge (T_1 = u)] \, du; \tag{6.21}$$

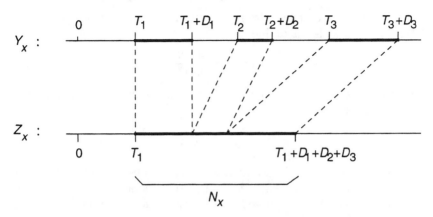

Figure 6.4 Approximation to the random present value of an LTC annuity.

note that this approximation involves an implicit safety loading (see (6.19)).

Moreover, let us define for any u:

$$n(x, u) = E[N_x | (S(0) = 1) \wedge (T_1 = u)]; \tag{6.22}$$

it is well known that

$$\bar{a}_{E(R|H)]} \geq E[\bar{a}_R] | H]; \tag{6.23}$$

whatever the random duration R and the conditioning event H are. Hence, introducing a further safety loading, we can assume:

$$\mathscr{B}_1^{LTC}(0, \infty) \cong b \int_0^{+\infty} P_{\underline{11}}(0, u)\mu_{13}(u)v^u \bar{a}_{n(x,u)]} \, du. \tag{6.24}$$

Formula (6.24) requires the following data:

- the probability of remaining healthy, $P_{\underline{11}}(0, u)$;
- the probability of disablement, $\mu_{13}(u) \, du$;
- the expected duration of disability, $n(x, u)$, as a function of u.

Furthermore, formula (6.2) can be transformed in order to be supported by available statistical data. Let us define:

$$\phi(x, u) \, du = P_{\underline{11}}(0, u)\mu_{13}(u) \, du. \tag{6.25}$$

Hence, $\phi(x, u) \, du$ is the probability that the first LTC claim occurs between ages $x + u$ and $x + u + du$; it follows that

$$\Phi(x) = \int_0^{+\infty} \phi(x, u) \, du \tag{6.26}$$

is the probability that the (first) exit from the healthy state is due to an LTC claim. Now, let us define

$$\psi(x, u) = \frac{\phi(x, u)}{\Phi(x)} \tag{6.27}$$

which represents the probability density function of the random time T_1. Hence, we can write:

$$\mathcal{B}_1^{\text{LTC}}(0, \infty) \cong b\Phi(x) \int_0^{+\infty} \psi(x, u)v^u \bar{a}_{n(x,u)]} \, du. \tag{6.28}$$

The $\Phi(x)$ values can be estimated by the number of persons, among those of age x at policy issue, requiring LTC benefits sooner or later; the estimate of $\psi(x, u)$ can be based on the statistical distribution of the age at the time of first LTC claim.

Some comments can help in interpreting the practical implications as well as the error involved using the approximation above described. As regards the approximation error involved by formula (6.21) it is self-evident that it decreases as the probabilities of recovery decrease. Conversely, to assess the effect of formula (6.24) the probability distribution of N_x, conditional on $(S(0) = 1) \wedge (T_1 = u)$, should be analysed.

As far as the effectiveness of the approximation method is concerned, it should be pointed out that it strongly depends on the type of data available. In fact, the 'exact' calculation of $\mathcal{B}_1^{\text{LTC}}(0, \infty)$ requires the estimation of four intensity functions, $\mu_{13}(u)$, $\mu_{31}(u)$, $\mu_{12}(u)$ and $\mu_{32}(u)$, and the use of a formula (allowing for recovery) like equation (1.157) (see section 1.12 in Chapter 1; note that state 3 must now be considered in place of state 2). So, when data are available for a straightforward estimation of the functions $\Phi(x)$, $\psi(x, u)$ and $n(x, u)$, the approximation method above described might be preferred.

6.3.3 Enhanced annuity

This LTC product consists of a life annuity sold to an elderly person at the time of entering into a nursing home, or to current residents of nursing homes. Of course this product requires the payment of a single premium. The relevant two-state model is depicted in Fig. 6.5.

Let $\mu_{32}(t)$ denote the force of mortality of the insured, i.e. of a person LTC claiming. Of course $\mu_{32}(t) > \mu(t)$, if $\mu(t)$ denotes the force of mortality adopted for pricing a standard life annuity. For this cover we have:

$$\mathcal{B}_3^{\text{LTC}}(0, \infty) = b_3 \bar{a}_{33}(0, \infty) = b_3 \int_0^{+\infty} P_{33}(0, t)v^t \, dt$$

$$= b_3 \int_0^{+\infty} v^t \exp\left[-\int_0^t \mu_{32}(z) \, dz\right] dt. \tag{6.29}$$

Figure 6.5 A two-state model for the enhanced annuity.

The actuarial value of a standard life annuity is:

$$\bar{b}\bar{a}(0,\infty) = b \int_0^{+\infty} v^t \exp\left[-\int_0^t \mu(z)\,dz\right]dt. \tag{6.30}$$

For a given single premium π, let b and b_3 be such that

$$\pi = b_3\bar{a}_{33}(0,\infty) = b\bar{a}(0,\infty). \tag{6.31}$$

From $\mu_{32}(t) > \mu(t)$ it follows $\bar{a}_{33}(0,\infty) < \bar{a}(0,\infty)$ and then the enhancement is given by $b_3 - b > 0$.

The formula for the prospective reserve is straightforward.

6.3.4 LTC annuity as a rider benefit

Let us consider the multiple state model depicted in Fig. 6.6. Thus, state 3 is the LTC state. Let us consider a whole life assurance with death benefit $c_{12}(t) = c$. Let b_3 denote the constant rate of the LTC annuity. Assume

$$b_3 = c/r \tag{6.32}$$

where r denotes a given number of years, representing the maximum number of years of LTC annuity payment.

For an LTC claiming policyholder, let us define the death benefit function, $c_{32}(t)$, as follows:

$$c_{32}(t) = c - b_3 \min(z,r) \tag{6.33}$$

where z denotes the time spent in state 3 up to time t. Thus, the LTC rider benefit consists in the pre-payment of the death benefit, in the form of an annuity.

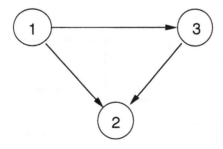

Figure 6.6 A three-state model for an LTC annuity with one state of disability, no recovery admitted.

For brevity, we restrict our attention to premium calculation only. The actuarial value of the benefits at policy issue is given by:

$$\mathscr{B}_1^{\mathrm{LTC}}(0,\infty) = b_3 \int_0^{+\infty} P_{\underline{11}}(0,u)\mu_{13}(u)v^u \bar{a}_{\underline{33}}(u,u+r)\,\mathrm{d}u + c\bar{A}_{112}(0,\infty)$$

$$+ \int_0^{+\infty} P_{\underline{11}}(0,u)\mu_{13}(u)v^u$$

$$\times \int_0^r (c-b_3 z)v^z P_{33}(u,u+z)\mu_{32}(u+z)\,\mathrm{d}z\,\mathrm{d}u. \tag{6.34}$$

The constant premium rate p, assuming that continuous premiums are payable for m years while the insured is healthy, is determined by the equivalence principle:

$$p\bar{a}_{\underline{11}}(0,m) = \mathscr{B}_1^{\mathrm{LTC}}(0,\infty). \tag{6.35}$$

6.3.5 Enhanced pension

Assume that at the time of retirement the amount S is available to the insured (for example, a member of an occupational scheme; in this case S may be the amount accrued during the service period). According to the 'enhanced pension' policy design, at the time of retirement the insured can choose between (see Fig. 6.7):

1. a straight life annuity (**basic pension**), at a rate b; and
2. • an annuity while he is in state 1, at a rate b_1, $b_1 < b$;
   • an LTC annuity (**enhanced pension**), at a rate b_3, $b_3 > b$.

Using S as a single premium, the rate b of the basic pension (choice 1) is determined as follows:

$$S = b[\bar{a}_{\underline{11}}(0,\infty) + \bar{a}_{13}(0,\infty)]. \tag{6.36}$$

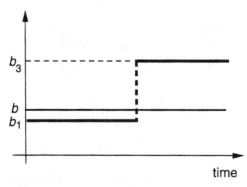

Figure 6.7 Enhanced pension arrangement.

The actuarial value of the benefits under choice 2 is given by:

$$\mathcal{B}_1^{\text{LTC}}(0,\infty) = b_1 \bar{a}_{\underline{11}}(0,\infty) + b_3 \bar{a}_{13}(0,\infty). \tag{6.37}$$

The rates b_1 and b_3 must be determined requiring that:

$$\mathcal{B}_1^{\text{LTC}}(0,\infty) = S \tag{6.38}$$

whence

$$b = \frac{b_1 \bar{a}_{\underline{11}}(0,\infty) + b_3 \bar{a}_{13}(0,\infty)}{\bar{a}_{\underline{11}}(0,\infty) + \bar{a}_{13}(0,\infty)}; \tag{6.39}$$

thus b is the weighted arithmetic mean of b_1 and b_3.

The prospective reserves (relating to choice 2) are given by:

$$\bar{V}_1(t) = b_1 \bar{a}_{\underline{11}}(t,\infty) + b_3 \bar{a}_{13}(t,\infty) \tag{6.40}$$

$$\bar{V}_3(t) = b_3 \bar{a}_{\underline{33}}(t,\infty). \tag{6.41}$$

6.3.6 An insurance package

Let us consider the three-state model represented by Fig. 6.6 and the following set of benefits:

1. a deferred life annuity at a constant rate b_1, paid while the insured is in state 1; let n denote the deferment period;
2. an LTC annuity at a constant rate b_3, paid while the insured is in state 3;
3. a constant death benefit $c_{12}(t) = c_{32}(t) = c$.

For brevity, we only consider the method of premium calculation. The actuarial value of the benefits is given by:

$$\mathcal{B}_1^{\text{LTC}}(0,\infty) = b_1 P_{11}(0,n) v^n \bar{a}_{\underline{11}}(n,\infty) + b_3 \bar{a}_{13}(0,\infty)$$
$$+ c[\bar{A}_{112}(0,\infty) + \bar{A}_{132}(0,\infty)]. \tag{6.42}$$

The constant premium rate p, assuming that continuous premiums are payable for m years ($m \le n$) while the insured is healthy, is determined by the equivalence principle:

$$p\bar{a}_{\underline{11}}(0,m) = \mathcal{B}_1^{\text{LTC}}(0,\infty). \tag{6.43}$$

The benefit arrangement can be modified in several ways. In particular, for given values of b and c, let us consider the policy design defined as follows:

1. $b_1 = b$;
2. $b_3 = b$;

3. $c_{12}(t) = c_{32}(t) = \max[0, c - b(z_1 + z_3)] = c - \min[c, b(z_1 + z_3)]$, where:

$\qquad z_1 = $ time spent in state 1 from time n up to time t;

$\qquad z_3 = $ time spent in state 3 up to time t.

As far as the death benefit is concerned, the relevant actuarial value can be split up as follows:

1. the actuarial value of the death benefit paid before time n if the insured dies while in the 'healthy' state:

$$c\bar{A}_{112}(0, n);$$

2. the actuarial value of the death benefit paid after time n if the insured dies while in the 'healthy' state:

$$P_{\underline{11}}(0, n)v^n \int_0^{c/b} (c - bu)v^u P_{\underline{11}}(n, n+u)\mu_{12}(n+u)\, du;$$

3. the actuarial value of the death benefit paid to the insured who entered the LTC state before time n:

$$\int_0^n \left[P_{\underline{11}}(0, t)\mu_{13}(t)v^t \int_0^{c/b} (c - bu)v^u P_{\underline{33}}(t, t+u)\mu_{32}(t+u)\, du \right] dt;$$

4. the actuarial value of the death benefit paid to the insured who entered the LTC state after time n:

$$P_{\underline{11}}(0, n)v^n \int_0^{c/b} (c - bu)v^u P_{13}(n, n+u)\mu_{32}(n+u)\, du.$$

Hence the actuarial value of the benefits is given by:

$$\mathscr{B}_1^{\text{LTC}}(0, \infty) = b[P_{\underline{11}}(0, n)v^n\bar{a}_{\underline{11}}(n, \infty) + \bar{a}_{13}(0, \infty)]$$

$$+ c\bar{A}_{112}(0, n) + \int_0^n \left[P_{\underline{11}}(0, t)\mu_{13}(t)v^t \right.$$

$$\left. \times \int_0^{c/b} (c - bu)v^u P_{\underline{33}}(t, t+u)\mu_{32}(t+u)\, du \right] dt$$

$$+ P_{\underline{11}}(0, n)v^n \int_0^{c/b} (c - bu)v^u [P_{\underline{11}}(n, n+u)\mu_{12}(n+u)$$

$$+ P_{13}(n, n+u)\mu_{32}(n+u)]\, du. \qquad (6.44)$$

The equation for the level premium is then straightforward (see (6.43)).

6.4 THE TIME-DISCRETE APPROACH

In this section we describe a time-discrete Markov model for LTC benefits. A similar model for disability benefits has been already presented in

Table 6.1 Transition probabilities

	a	i'	i''	d
a	p_y^{aa}	$p_y^{ai'}$	$p_y^{ai''}$	q_y^a
i'	0	$p_y^{i'i'}$	$p_y^{i'i''}$	$q_y^{i'}$
i''	0	0	$p_y^{i''i''}$	$q_y^{i''}$
d	0	0	0	1

Chapter 2, sections 2.1 to 2.3, and in Chapter 3, section 3.2.7. So we can now restrict our attention to some aspects strictly concerning LTC covers.

As regards notation, we use the symbols adopted in Chapter 3 (see in particular section 3.2.7), since these symbols are commonly used in practical time-discrete implementations. The same notation will be adopted in section 6.5, in which some numerical examples are discussed.

6.4.1 Probabilities for time-discrete models

Let x denote the (integer) age at policy entry. We now consider the process $\{S(y); y = x, x + 1, \ldots\}$ assuming that it is a time-discrete inhomogeneous Markov chain. Consistently with the notation used for disability benefits, we now denote the states as follows:

$$a = \text{active, i.e. healthy;}$$
$$i' = \text{frailty at level I;}$$
$$i'' = \text{frailty at level II;}$$
$$d = \text{dead.}$$

Table 6.1 is the matrix of the transition probabilities related with the model depicted in Fig. 6.2. Since no recovery is admitted, the matrix is upper triangular.

In terms of $S(y)$, we have for example:

$$p_y^{aa} = \Pr\{S(y + 1) = a \,|\, S(y) = a\}$$
$$p_y^{ai'} = \Pr\{S(y + 1) = i' \,|\, S(y) = a\}$$
$$q_y^a = \Pr\{S(y + 1) = d \,|\, S(y) = a\}.$$

For $h = 1, 2, \ldots$, we have (Chapman–Kolmogorov relations):

$$_hp_y^{aa} = {}_{h-1}p_y^{aa}\, p_{y+h-1}^{aa}; \tag{6.45}$$

$$_hp_y^{ai'} = {}_{h-1}p_y^{ai'}\, p_{y+h-1}^{i'i'} + {}_{h-1}p_y^{aa}\, p_{y+h-1}^{ai'}; \tag{6.46}$$

$$_hp_y^{ai''} = {}_{h-1}p_y^{ai''}\, p_{y+h-1}^{i''i''} + {}_{h-1}p_y^{ai'}\, p_{y+h-1}^{i'i''} + {}_{h-1}p_y^{aa}\, p_{y+h-1}^{ai''}. \tag{6.47}$$

Moreover, the following relations hold:

$$_h p_y^{ai'} = \sum_{r=1}^{h} {}_{h-r}p_y^{aa} p_{y+h-r}^{ai'} \prod_{g=1}^{r-1} p_{y+h-r+g}^{i'i'};$$ (6.48)

$$_h p_y^{ai''} = \sum_{r=1}^{h} \left[{}_{h-r}p_y^{aa} p_{y+h-r}^{ai''} \prod_{g=1}^{r-1} p_{y+h-r+g}^{i''i''} + {}_{h-r}p_y^{ai'} p_{y+h-r}^{i'i''} \prod_{g=1}^{r-1} p_{y+h-r+g}^{i''i''} \right].$$ (6.49)

6.4.2 Premiums and reserves for stand-alone annuities

Let b', b'' denote the annual payments related to LTC states i' and i'' respectively. Assume that the benefits are paid at policy anniversaries while the insured is claiming LTC benefits. The actuarial value of the LTC annuities at time 0 (then conditional on $S(x) = a$) and hence the single premium are given by:

$$\mathcal{B}_a^{LTC}(0, \infty) = \sum_{h=1}^{+\infty} (b' \, _h p_x^{ai'} + b'' \, _h p_x^{ai''}) v^h$$

$$= \sum_{h=1}^{+\infty} \left[b' \sum_{r=1}^{h} {}_{h-r}p_x^{aa} p_{x+h-r}^{ai'} \prod_{g=1}^{r-1} p_{x+h-r+g}^{i'i'} \right.$$

$$+ b'' \sum_{r=1}^{h} {}_{h-r}p_x^{aa} p_{x+h-r}^{ai''} \prod_{g=1}^{r-1} p_{x+h-r+g}^{i''i''}$$

$$\left. + b'' \sum_{r=1}^{h} {}_{h-r}p_x^{ai'} p_{x+h-r}^{i'i''} \prod_{g=1}^{r-1} p_{x+h-r+g}^{i''i''} \right] v^h.$$ (6.50)

Finally, with obvious definitions for $\ddot{a}_{x+j}^{i'}$ and $\ddot{a}_{x+j}^{i''}$ and after some manipulations, we get:

$$\mathcal{B}_a^{LTC}(0, \infty) = b' \sum_{j=1}^{+\infty} {}_{j-1}p_x^{aa} p_{x+j-1}^{ai'} v^j \ddot{a}_{x+j}^{i'} + b'' \sum_{j=1}^{+\infty} {}_{j-1}p_x^{aa} p_{x+j-1}^{ai''} v^j \ddot{a}_{x+j}^{i''}$$

$$+ b'' \sum_{j=2}^{+\infty} {}_{j-1}p_x^{ai'} p_{x+j-1}^{i'i''} v^j \ddot{a}_{x+j}^{i''}.$$ (6.51)

It is interesting to observe that (6.51) is a discrete 'inception-annuity' formula, which can be considered an extension of the well-known formulae used for premiums and reserves in disability insurance (see, for some examples, section 3.4.1).

The definition of the actuarial values of the LTC annuity benefit can be easily generalized to any integer time t, $t = 0, 1, 2, \ldots$, and relevant states

$S(x + t)$. We find:

$$\mathcal{B}_a^{\text{LTC}}(t, \infty) = b' \sum_{j=1}^{+\infty} {}_{j-1}p_{x+t}^{aa} p_{x+t+j-1}^{ai'} v^j \ddot{a}_{x+t+j}^{i'}$$

$$+ b'' \sum_{j=1}^{+\infty} {}_{j-1}p_{x+t}^{aa} p_{x+t+j-1}^{ai''} v^j \ddot{a}_{x+t+j}^{i''}$$

$$+ b'' \sum_{j=2}^{+\infty} {}_{j-1}p_{x+t}^{ai'} p_{x+t+j-1}^{i'i''} v^j \ddot{a}_{x+t+j}^{i''}; \qquad (6.52)$$

$$\mathcal{B}_{i'}^{\text{LTC}}(t, \infty) = b' \ddot{a}_{x+t}^{i'} + b'' \sum_{j=1}^{+\infty} {}_{j-1}p_{x+t}^{i'i'} p_{x+t+j-1}^{i'i''} v^j \ddot{a}_{x+t+j}^{i''}. \qquad (6.53)$$

$$\mathcal{B}_{i''}^{\text{LTC}}(t, \infty) = b'' \ddot{a}_{x+t}^{i''}. \qquad (6.54)$$

Assume that annual level premiums are paid for m years while the insured is healthy. Let p denote the annual premium. The equivalence principle implies that:

$$p\ddot{a}_{x:m|}^{aa} = \mathcal{B}_a^{\text{LTC}}(0, \infty) \qquad (6.55)$$

where

$$\ddot{a}_{x:m|}^{aa} = \sum_{j=0}^{m-1} v^j \, {}_j p_x^{aa}. \qquad (6.56)$$

Three reserves must be defined, relating to the three (strictly) transient states. We shall restrict our attention to reserves at integer times. The healthy reserve (hence conditional on $S(x + t) = a$) is given by:

$$V_t^a = \mathcal{B}_a^{\text{LTC}}(t, \infty) - p\ddot{a}_{x+t:m-t|}^{aa} \qquad \text{if } t < m \qquad (6.57)$$

$$V_t^a = \mathcal{B}_a^{\text{LTC}}(t, \infty) \qquad \text{if } t \geq m. \qquad (6.58)$$

The reserves relating to the two LTC states (hence conditional on $S(x + t) = i'$ and $S(x + t) = i''$, respectively) are given by:

$$V_t^{i'} = \mathcal{B}_{i'}^{\text{LTC}}(t, \infty) \qquad (6.59)$$

$$V_t^{i''} = \mathcal{B}_{i''}^{\text{LTC}}(t, \infty). \qquad (6.60)$$

6.4.3 Premiums for an LTC annuity as a rider benefit

Let c denote the death benefit for a non-LTC claiming insured. Let b denote the annual payment related to the LTC state (see Fig. 6.6, with $1 = a$, $2 = d$, $3 = i$). Assume (as in the time-continuous case – see section 6.3.4):

$$b = c/r \tag{6.61}$$

where r denotes the maximum number of years of LTC annuity payment. For an insured who is LTC claiming, let us define the death benefit as follows:

$$c_i(t) = c - b \min(h, r) \tag{6.62}$$

where h denotes the number of LTC annuity payments up to time t.

For brevity, we concentrate our attention only on premium calculation. The actuarial value of the benefits at policy issue (hence with $S(x) = a$) is given by:

$$\mathcal{B}_a^{\text{LTC}}(0, \infty) = c \sum_{j=1}^{+\infty} v^j{}_{j-1}p_x^{aa} q_{x+j-1}^a + \sum_{j=1}^{+\infty} \left[{}_{j-1}p_x^{aa} p_{x+j-1}^{ai} v^j \right.$$

$$\left. \times \left(b\ddot{a}_{x+j:\overline{r}|}^i + \sum_{h=1}^r (c - hb){}_{h-1}p_{x+j}^{ii} q_{x+j+h-1}^i v^h \right) \right]. \tag{6.63}$$

Assume that annual level premiums are paid for m years while the insured is healthy. Let p denote the annual premium. The equivalence principle implies that:

$$p\ddot{a}_{x:\overline{m}|}^{aa} = \mathcal{B}_a^{\text{LTC}}(0, \infty). \tag{6.64}$$

6.4.4 Premiums and reserves for the enhanced pension

Let b denote the annual payment relating to a straight life annuity (choice 1, see section 6.3.5). Let b_a and b_i respectively denote the annual payments relating to a life annuity paid while the insured is healthy and to a life annuity paid while he is claiming LTC benefits (choice 2). Assume that the amount S is available at the date of retirement. Using S as a single premium, the amount b is determined as follows:

$$S = b(\ddot{a}_x^{aa} + a_x^{ai}) \tag{6.65}$$

where

$$a_x^{ai} = \sum_{j=1}^{+\infty} {}_{j-1}p_x^{aa} p_{x+j-1}^{ai} v^j \ddot{a}_{x+j}^i. \tag{6.66}$$

It should be pointed out that in actuarial practice the expected present value $\ddot{a}_x^{aa} + a_x^{ai}$ is often approximated by the expected present value of a life annuity calculated by ignoring the known (healthy) state at age x, i.e. by:

$$\ddot{a}_x = \sum_{j=0}^{+\infty} v^j{}_j p_x. \tag{6.67}$$

Of course actuarial values such as $\ddot{a}_x$ and probabilities such as $_jp_x$ are not well defined within the context of the three-state Markov model. Moreover, probabilities $_{j-1}p_x^{aa}$ in (6.66) are commonly replaced by probabilities $_{j-1}p_x$ (see for example section 3.4.1 for a discussion in respect of disability insurance).

Amounts b_a and b_i are related to amount S as follows:

$$S = b_a \ddot{a}_x^{aa} + b_i a_x^{ai} \tag{6.68}$$

whence

$$b = \frac{b_a \ddot{a}_x^{aa} + b_i a_x^{ai}}{\ddot{a}_x^{aa} + a_x^{ai}}. \tag{6.69}$$

The prospective reserves relating to choice 2 are given by:

$$V_t^a = b_a \ddot{a}_{x+t}^{aa} + b_i a_{x+t}^{ai} \tag{6.70}$$

$$V_t^i = b_i \ddot{a}_{x+t}^i. \tag{6.71}$$

The approximations mentioned above then lead to

$$V_t^a = b_a \ddot{a}_{x+t} + (b_i - b_a) a_{x+t}^{ai}. \tag{6.72}$$

6.5 SOME NUMERICAL EXAMPLES

In this section we discuss some numerical examples concerning premiums and reserves for LTC annuities. Results are presented for convenience in graphical form. Calculations are based on a time-discrete model (see section 6.4). Examples concern stand-alone annuities, LTC annuities as rider benefits, and enhanced pensions.

The following force of mortality has been assumed in order to represent the overall mortality (see section 3.3.1):

$$\mu_y = 0.0005 + 10^{0.038y - 4.12}. \tag{6.73}$$

From equation (6.73) the survival function has been derived and hence the annual survival and death probabilities p_y, q_y, $p_y + q_y = 1$.

For pricing as well as reserving, present values have been calculated using the annual discount factor $v = 1.04^{-1}$.

6.5.1 Premiums and reserves for stand-alone annuities

As far as frailty is concerned, survey data from the UK Office of Population Censuses and Surveys (OPCS), regarding the prevalence of disability among adults (see Martin *et al.*, 1988), have been used and transformed according to the procedure suggested by Gatenby (1991).

For the mortality of LTC claiming policyholders we have assumed:

$$q_y^{i'} = (1 + \eta_1)q_y; \quad \eta_1 \geq 0 \tag{6.74}$$

$$q_y^{i''} = (1 + \eta_2)q_y; \quad \eta_2 \geq \eta_1 \tag{6.75}$$

with q_y derived from equation (6.73). Furthermore we have assumed:

$$p_y^{aa} = p_y. \tag{6.76}$$

Note that this assumption reflects the hypothesis usually adopted in pricing disability annuities; in that context assumption (6.76) is commonly accepted especially when annual premiums are concerned (see, for example, sections 3.4.1 and 3.4.5). Finally we have assumed

$$p_y^{ai''} = p_y^{i'i''}. \tag{6.77}$$

Figure 6.8 illustrates the behaviour of probabilities $p_y^{ai'}$ and $p_y^{ai''}$ derived from the OPCS male data. All the following calculations have been performed using the male data.

First, we analyse the single premium as a function of age at entry (see Fig. 6.9). The behaviour of this function can be interpreted in terms of the joint effect of the probabilities of becoming LTC claiming and the probabilities of death, which both increase as attained age increases. Annual premiums (see Fig. 6.10) are assumed to be payable up to age 65 (which can be interpreted as the retirement age).

Figure 6.11 illustrates the reserve for the healthy state, in the case of a single premium and of annual premiums, as a function of attained age. We note the particular profile of the reserve during the premium paying period in the latter case. LTC reserves are depicted in Figs 6.12

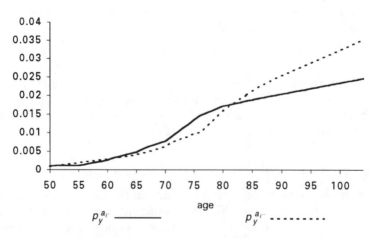

Figure 6.8 Probabilities from OPCS data.

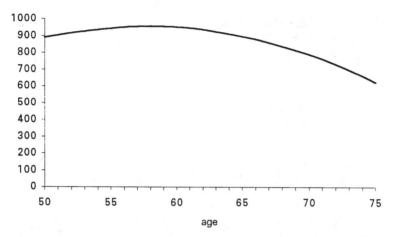

Figure 6.9 Single premium as a function of the age at entry, $b' = 600$, $b'' = 1000$, $\eta_1 = 0.05$, $\eta_2 = 0.10$.

and 6.13. Note that the LTC reserves are functions of the attained age only, as a non-inception-select model is being used.

The single premium as a function of the mortality of the LTC claiming policyholders (see formulae (6.74) and (6.75)) is illustrated in Fig. 6.14. Note that for all ages at entry the single premium decreases as the assumed extra level of mortality (the parameter η_1) increases. This can be easily understood, since the benefits consist of annuities.

Figure 6.15 illustrates, for various ages at entry, the behaviour of the single premium as a function of the parameter α which represents the ratio of the benefit amount b' to the amount b''. The single premium is a

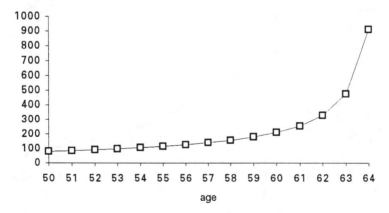

Figure 6.10 Annual premium, payable for $65 - x$ years, as a function of the age x at entry, $b' = 600$, $b'' = 1000$, $\eta_1 = 0.05$, $\eta_2 = 0.10$.

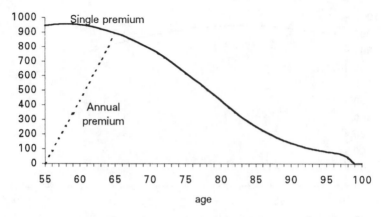

Figure 6.11 Reserve for the healthy state: age at entry $x = 55$, $b' = 600$, $b'' = 1000$, $\eta_1 = 0.05$, $\eta_2 = 0.10$.

linear function of this parameter, as it is easy to check by inspection of the relevant formula (see (6.51)).

Finally a sensitivity analysis has been performed, in order to represent the single premium as a function of the probabilities of becoming LTC claiming. To this purpose, we have used the probabilities $\gamma p_y^{ai'}$, $\gamma p_y^{ai''}$ and $\gamma p_y^{i'i''}$ (where, as before, $p_y^{ai''} = p_y^{i'i''}$). Figure 6.16 illustrates some of these results. Note that, for any given age at entry, the single premium is an increasing function of γ. However, the effect of γ diminishes as the age at entry increases, since the expected duration of the LTC annuities decreases. It must be stressed that this sensitivity analysis is important because data are scanty.

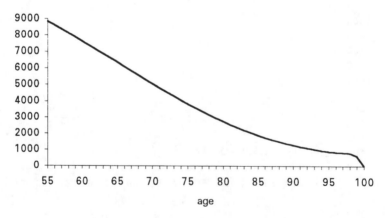

Figure 6.12 Reserve for the LTC state i': age at entry $x = 55$, $b' = 600$, $b'' = 1000$, $\eta_1 = 0.05$, $\eta_2 = 0.10$.

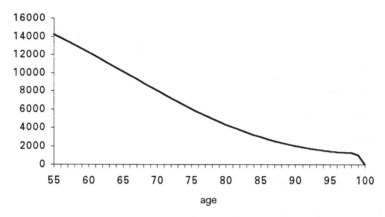

Figure 6.13 Reserve for the LTC state i'': age at entry $x = 55$, $b'' = 1000$, $\eta_2 = 0.10$.

6.5.2 Premiums and reserves for an LTC annuity as a rider benefit

In this case just one LTC state is taken into account (see Fig. 6.6, with state $3 = i$). Consistently with the three-state structure we have assumed

$$p_x^{ai} = p_x^{ai'} + p_x^{ai''}. \tag{6.78}$$

The amount of the death benefit depends on the state (non-LTC versus LTC claiming) at the time of death and, in particular, on the number of LTC annuity payments up to the time of death. The benefit design obviously affects the single premium (see formula (6.63)). Then a distinction has to be made between the probabilities of remaining healthy and the probabilities of remaining in the LTC state, the two events determining the state at the time of death and then the benefits. Hence assumption

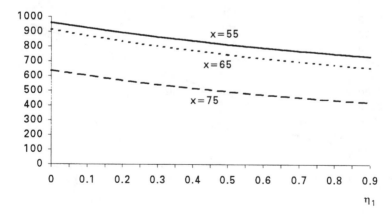

Figure 6.14 Single premium as a function of η_1 and η_2, with $\eta_2 = 1.05(1 + \eta_1) - 1$, for various ages at entry, $b' = 600$, $b'' = 1000$.

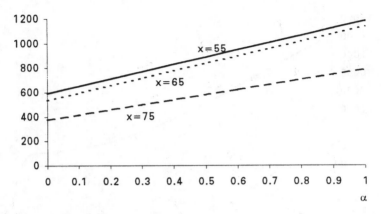

Figure 6.15 Single premium as a function of α, with η_2, with $b' = \alpha b''$, for various ages at entry, $b'' = 1000$, $\eta_1 = 0.05$, $\eta_2 = 0.10$.

(6.76) in section 6.5.1 cannot be accepted. Therefore, for this (and the following) example we assume:

$$p_x^{aa} = p_x - p_x^{ai} \tag{6.79}$$

with p_x^{ai} given by (6.78). As far as the mortality of LTC-claiming insureds is concerned, we assume (as in section 6.5.1):

$$q_y^i = (1 + \eta)q_y; \quad \eta \geq 0. \tag{6.80}$$

We first analyse the single premium as a function of the age at entry, assuming that the annual LTC benefit is equal to one-fifth of the sum assured (the death benefit in the case of death when non-claiming LTC

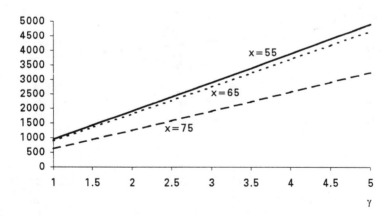

Figure 6.16 Single premium as a function of γ, and then as a function of probabilities $\gamma p_y^{ai'}$, $\gamma p_y^{ai''}$, for various ages at entry, $b' = 600$, $b'' = 1000$, $\eta_1 = 0.05$, $\eta_2 = 0.10$.

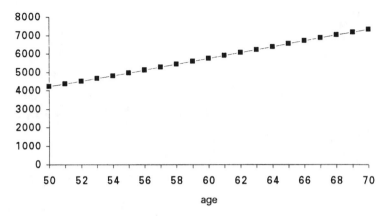

Figure 6.17 Single premium as a function of age at entry, $c = 10\,000$, $r = 5$, $\eta = 0.05$.

benefits). Results are presented in Fig. 6.17. The single premium increases as the age at entry increases, because of the higher probabilities of both death and claiming LTC.

Figure 6.18 illustrates the behaviour of the single premium as a function of the maximum number r of annuity payments and hence of the amount of the annual LTC payment. Note that the 'pre-payment' effect decreases as r increases, and hence also the single premium decreases as r increases.

Finally, the same results are presented in Fig. 6.19 in terms of the ratio between the single premium of this cover and the single premium for a traditional whole life assurance.

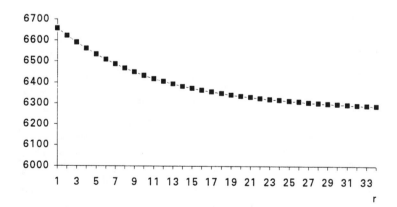

Figure 6.18 Single premium as a function of r, $c = 10\,000$, $x = 65$, $\eta = 0.05$.

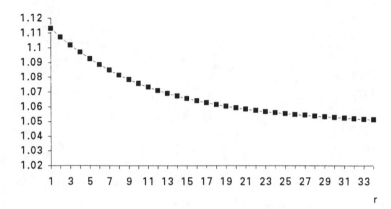

Figure 6.19 Single premium as a function of r compared with the single premium of a conventional whole-life assurance, $c = 10\,000$, $x = 65$, $\eta = 0.05$.

6.5.3 Premiums and reserves for the enhanced pension

Also in this case just one LTC state is taken into account. Numerical calculations have been performed assuming the hypotheses adopted in section 6.5.2 (see equations (6.78), (6.79) and (6.80) in particular).

It seems particularly interesting to analyse the amount b_i as a function of b_a, once the single premium (and hence the amount b) has been assigned. From equation (6.68) it clearly follows that, for any age x at entry, b_i is a linear function of b_a. For example, with $x = 65$ we find $b_i = 3438$ if $b = 1000$ and $b_a = 700$. For this case, Figs 6.20 and 6.21 illustrate the reserves in respect of the healthy state and the LTC state respectively.

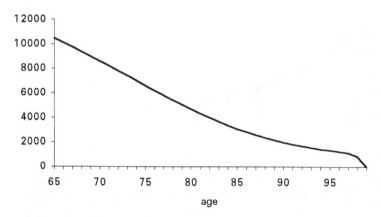

Figure 6.20 Reserve for the healthy state: age at entry $x = 65$, $b = 1000$, $b_a = 700$, $b_i = 3438$, $\eta = 0.05$.

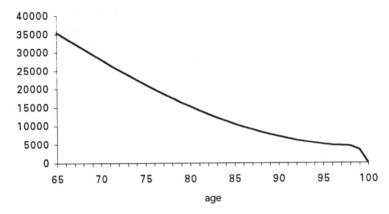

Figure 6.21 Reserve for the LTC state: age at entry $x = 65$, $b = 1000$, $b_a = 700$, $b_i = 3438$, $\eta = 0.05$.

6.6 REFERENCES AND SUGGESTIONS FOR FURTHER READING

Actuarial literature on LTC products is still rather scanty, whereas a number of articles and surveys describe ageing population problems and the need of long term care as well as LTC insurance policies.

In this section we first suggest some papers describing LTC products and related benefit design aspects. To this purpose the reader is referred to Bague (1991), Benjamin (1992), Booth (1996), Cowley (1992), Eagles (1992), Fennell (1991), Heistermann *et al.* (1990), Maynard (1991), Münchener Rück (1992), Nuttall (1992), Nuttall *et al.* (1994), Pollard (1995), PPP Lifetime Care (1996), Propp (1992), Walker (1992). A statistical analysis of the active life expectancies is presented by Beekman and Frye (1991).

A number of authors have considered the actuarial and statistical aspects of LTC insurance. Gatenby (1991) deals with the problem of deriving inception rates from prevalence data relating to a population. In Levikson and Mizrahi (1994) several methods describing the changes in the LTC need over time are analysed and some methods for pricing LTC products are proposed. Olivieri (1996) studies, in a time-continuous context, problems related with the technical bases for pricing and reserving LTC annuities and with the so-called 'safe-side' requirements. Pitacco (1994) analyses the derivation from a general multiple-state Markov model of particular, practical premium calculation schemes for a range of products. In Beekman (1990) approximation formulae for the actuarial value of an LTC annuity are proposed. The handbook of Münchener Rück (1992) is also useful for a discussion of actuarial and statistical aspects.

US statistical data concerning LTC are presented by Jones (1992) and Corliss *et al.* (1996) as well as in Report of the Long-Term-Care Experience Committee (1992). UK data are presented by Martin *et al.* (1988).

A number of papers deal with statistical and actuarial problems in related fields, such as geriatric wards and CCRCs (Continuing Care Retirement Communities). In particular readers are referred to Jones (1995, 1996, 1997a, 1997b), Jones and Willmot (1993), Meredith (1973) and Winklevoss and Powell (1984).

7

Actuarial models for AIDS

7.1 INTRODUCTION

The objective of this chapter is to consider some of the multiple state models that have been proposed in the actuarial literature for representing the spread of the AIDS epidemic with a view to considering the impact of AIDS on insurance, pensions and social security. Our aim is not to provide a comprehensive review of the models that have been presented in the epidemiological and statistical literature – for such a guide, readers are referred to the review papers of Gail and Brookmeyer (1988) or Isham (1988) for a statistical perspective. (It should be noted that the literature is both large and expanding, and so any such review is likely to become out of date fairly quickly.)

7.1.1 Why model?

A number of mathematical models of transmission of infection have been proposed for representing the spread of the AIDS epidemic. What is the point of attempting to model the spread of this epidemic?

From an epidemiological viewpoint, the most important purpose of constructing mathematical models is to provide a means of gaining an understanding about the real transmission dynamics of the infection and, in particular, of learning which features are likely to have a substantial influence on the course of the epidemic and what sort of effects are to be expected. It would thus be possible to investigate how changes in the various assumptions and parameter values of the models would affect the course of the epidemic. This understanding of the properties of the mathematical models and their sensitivities can then help to clarify how individuals, policy-makers and society should react to the epidemic (e.g. what behavioural changes should take place, what intervention strategies should be investigated) and how the consequences of the epidemic could be better managed (Isham, 1988). Thus, the whole process of modelling and data collection is iterative, with the act of modelling helping to structure our thoughts about the spread of infection and provide a framework within which to consider questions of the form: what would happen if A were changed to B? The modelling process then provides a guide to the sorts of data that should be collected to make better information available about the epidemic to society as a whole. From an actuarial point of view, such modelling is of vital importance in investigating the implications of the AIDS epidemic for insurance (life and disability) and pensions, in terms specifically of the effects on policy design, underwriting, pricing and reserving.

7.1.2 General points of epidemiology

It is not our purpose to provide a full discussion of the epidemiology of HIV infection and AIDS. (Interested readers are referred, for example, to Daykin, 1990.)

It is worth noting the following definitions (taken from Isham, 1988) relating to the course of AIDS in a particular susceptible individual (Fig. 7.1 provides an illustration of the corresponding stages). Three time periods are of importance: the **latent period**, the **infectious period** and the **incubation period**.

The latent period extends from the time of infection until the individual becomes infectious. This is followed by an infectious period during which the individual is termed an infective and can pass on the disease to

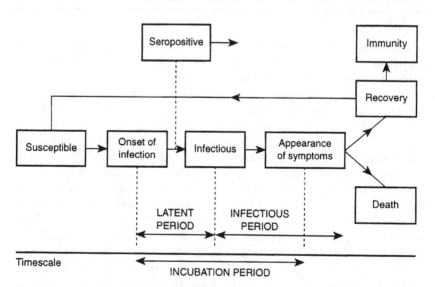

Figure 7.1 Stages for an infectious disease.

susceptibles. The incubation period is the period from infection until overt symptoms appear.

The individual infected with HIV is described as being **seropositive** when antibodies are normally detectable in the blood. This tends to occur a few weeks after infection. Given the likely length of the incubation period, it is convenient in modelling the epidemic to assume that all those infected with the virus are seropositive.

The individual may in due course recover or die (from the infection or from other causes) and, if he/she recovers, he/she may be immune to the underlying disease or may return once more to the susceptible state. Based on the evidence then available, some models have avoided including a class of individuals who have recovered and are immune.

The incubation period may be very long, and it is still debatable as to whether all those infected with HIV will ultimately progress to AIDS, or whether only a fraction will.

Many models have made the convenient and simplifying assumption that, after AIDS is diagnosed, the individual concerned is effectively isolated and unable to infect further susceptibles. This would mean that the infectious period is wholly contained within the incubation period (Isham, 1988).

7.1.3 Different approaches to modelling and prediction

Three distinct approaches have been put forward in the literature for modelling and predicting the spread of HIV infection and AIDS. These are as follows.

1. Extrapolation over time by regression or associated methods of the number of reported or diagnosed cases of AIDS so far.
2. Multiple state models of the transmission of infection and progression of disease.
3. Back-projection methods.

We shall be considering models in group 2 only in the remainder of this chapter, and we shall see how these fit into the general structure of multiple state models described in earlier chapters of this book.

7.2 INSTITUTE OF ACTUARIES AIDS WORKING PARTY MODEL

In the late 1980s, the Institute of Actuaries in England established a Working Party to investigate the wider insurance implications of AIDS. The Working Party developed a Markov model of the transmission and spread of AIDS among (only) male homosexuals in the United Kingdom: a full description is provided by Wilkie (1988a, 1989). The model proposed is similar, in many respects, to other mathematical models that have been proposed for the transmission and spread of AIDS and involves several approximations and assumptions which are required by the paucity of data and the complexity of the disease transmission and progression: these are discussed in the following paragraphs.

The emphasis of actuarial models of HIV/AIDS has been to determine the impact of HIV and AIDS on life insurance underwriting, premium rating and reserving, and, to a lesser extent, on disability insurance and pension provision. In section 7.5, we shall return to the disability insurance models of Chapter 3 and consider suitable modifications to deal with the presence of AIDS.

In constructing the model, some important prerequisites were taken into account. One feature is that the model should be age-specific, in order to consider the progress of an individual of a given age and sex through future calendar years, to consider the longer term trend in transmission and to produce numerical results (although not necessarily by analytical means). Thus, equilibrium models would be of lesser interest. It is also important for the model to reflect the type of data that would normally be available to an insurance company, and to allow for normal age-specific mortality as well as the extra morbidity and mortality attributable to AIDS.

For the above reasons, the Working Party's model is age-specific and the resulting numerical complexity has meant that elements that depend on detailed assumptions about sexual behaviour have been excluded.

As a simplification, the model assumes that the only mode of transmission of HIV infection is through sexual activity among male homosexuals, and that all reported cases in the United Kingdom have been among male homosexuals. This simplification was reasonable because, by the late 1980s, over 80% of reported cases in the UK were among homosexuals. However, it ignores the separate transmission dynamics among haemophiliacs, drug users and heterosexual males and females that later have come to play a larger part in the total epidemic of HIV infection (Wilkie, 1989).

A further simplification is the assumption that those males described below as being 'at risk' of infection behave in the same manner at any one time, so that the model does not recognise that the chance of infection may depend on the frequency of sexual contact or frequency of change of sexual partner Wilkie, 1989).

The model considers males of any single age as constituting a 'cohort', and traces their experience independently of other cohorts. It is assumed that infection occurs from a contact between two individuals within a single age group. This assumption is somewhat artificial but, if infections between those of different age groups balance out, may be an adequate representation of reality (Wilkie, 1989).

The members of one cohort at age x may be in any one of the 11 states that are depicted in Fig. 7.2. Five of these are live states: 'clear', 'at risk', 'immune', 'positive' and 'sick' from AIDS. Six are dead states; these are

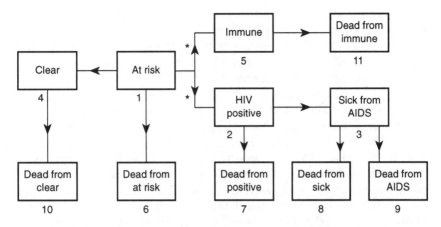

Figure 7.2 AIDS Working Party model (* denotes possible infection).

kept separate simply to identify the live state from which someone died. The dead states are: 'dead from clear', 'dead from at risk', 'dead from immune', 'dead from positive', 'dead from sick' (other than from AIDS) and 'dead from AIDS'. It may not be possible, in reality, to distinguish the last two categories, but deaths other than from AIDS among those who suffer from AIDS are comparatively small.

Those in the clear state are those whose assumed sexual behaviour is such that they have a zero risk of becoming infected with HIV. They form the 'normal' population for purposes of comparison. Those in the at-risk state are regarded as being at risk of acquiring HIV infection by reason of sexual contact with infected persons. Those in the immune state are assumed to have acquired the HIV and to be infectious, but to be immune from becoming sick from AIDS or dying from AIDS (Wilkie, 1989).

As noted on section 7.1.2, those in the positive state are not yet sick from AIDS; they are infectious and not immune. It is assumed that it is possible to distinguish between states 2 and 3, i.e. those who are HIV seropositive and those who are sick from AIDS. We note that there are several well-identified stages in the transition from HIV infection to death from AIDS (see section 7.4). Those who are suffering from AIDS may be very infectious, but it is possible that their sexual activity may be considerably reduced. The model makes it possible to choose whether those sick from AIDS are treated as infectious to further infections or not; in the version presented here, we assume that those sick from AIDS are *not* infectious (Wilkie, 1989).

It is assumed that the current age is part of the status, and that each transition intensity can vary with current age. In addition, since each age cohort is treated independently, each transition intensity can also be allowed to vary with calendar year, so that each cohort has its own set of transition intensities (Wilkie, 1989).

As in section 1.6, duration since entry to the states immune, positive and sick from AIDS is also assumed to be relevant to the corresponding transition intensities. This duration is denoted in each case by z.

Possible transitions are shown in Fig. 7.2. The entry state for an individual is assumed to be state 1: 'at risk'. Those in any of the live states may die, and those who are sick from AIDS may die from AIDS or from other causes. Those who are at risk may change their behaviour and become clear, for example, by giving up sexual activity altogether, or by restricting themselves to one equally monogamous partner. The model does not incorporate transfer from clear to at risk. Those who are at risk may become infected, and at that point are instantaneously split between the immune state and the positive state, in proportions that may depend on age (and on calendar year).

Those in the positive state may become sick from AIDS, if they do not die first. Infection is possible from the immunes and positives to the at risk, and the mechanism by which this is represented in the model is described below.

Conditional on being in state 1 at a starting age x_0, i.e. $S(x_0) = 1$, the probability of an individual being in state j at age x is denoted by $P_j(x)$ in a shorthand notation, where $j = 1, 4, 6, 7, 8, 9, 10, 11$ (equivalent to $P_{1j}(0, x - x_0)$ based on the notation of section 1.4). The corresponding probability for those in the positive, sick from AIDS or immune states at current age x and duration z is $P_j(x, z)$ where $j = 2, 3, 5$ respectively (equivalent to $P_{1j}(0, x - x_0, z)$ based on the notation of section 1.6). The total probability of being positive, sick from AIDS or immune at all durations is denoted by $P_j(x)$ for $j = 2, 3, 5$.

The transition intensity from state j to state k is represented by $\mu_{jk}(x)$, if it depends only on age x and by $\mu_{jk}(x, z)$ if it depends on both age x and duration z (in the current state). These are equivalent to $\mu_{jk}(x - x_0)$ and $\mu_{jk}(x - x_0, z)$ respectively based on the notation of sections 1.4 and 1.6.

7.2.1 Differential equations

The usual Kolmogorov differential equations describe the transitions between states, with the important exception of the transmission of infection. Following the development of Chapter 1, these equations are now presented:

$$\frac{d}{dx}P_4(x) = -\mu_{410}(x)P_4(x) + \mu_{14}(x)P_1(x) \tag{7.1}$$

$$\frac{d}{dx}P_1(x) = -\mu_{14}(x)P_1(x) - \mu_{16}(x)P_1(x) - T(x). \tag{7.2}$$

(In equation (7.2) we note that those at risk are diminished by transfers to the clear and dead states and by the transfers to the infectious states, represented by $T(x)$: see below for a further discussion.)

$$\frac{d}{dx}P_{10}(x) = \mu_{410}(x)P_4(x). \tag{7.3}$$

A similar equation to (7.3) applies to the differential for $P_j(x)$ where $j = 6$. For $j = 7, 8, 9, 11$ we have integro-differential equations, for example:

$$\frac{d}{dx}P_7(x) = \int_0^{x-x_o} P_2(x,z)\mu_{27}(x,z)\,dz.$$

We put $w = x - z$ so that w is the entry age to the current state. Then, for $z > 0$

$$\frac{d}{dz}P_5(w+z,z) = -\mu_{511}(w+z,z)P_5(w+z,z) \tag{7.4}$$

$$\frac{d}{dz}P_3(w+z,z) = -(\mu_{38}(w+z,z) + \mu_{39}(w+z,z))P_3(w+z,z) \tag{7.5}$$

$$\frac{d}{dz}P_2(w+z,z) = -\mu_{27}(w+z,z)P_2(w+z,z)$$

$$-\mu_{23}(w+z,z)P_2(w+z,z). \tag{7.6}$$

When duration z is 0, we have

$$\frac{d}{dx}P_3(x,0) = \int_0^{x-x_0} \mu_{23}(x,z)P_2(x,z)\,dz \tag{7.7}$$

$$\frac{d}{dx}P_5(x,0) = f(x)T(x) \tag{7.8}$$

and

$$\frac{d}{dx}P_2(x,0) = (1-f(x))T(x) \tag{7.9}$$

where $f(x)$ is the proportion of those newly infected at age x who are assumed to enter state 5, the immune state, and, as above, $T(x)$ is the density of those newly infected at age x.

It would be natural, in terms of the development of Chapter 1, to specify the right-hand sides of equations (7.8) and (7.9) to be

$$P_1(x)\mu_{15}(x)$$

and

$$P_1(x)\mu_{12}(x)$$

respectively, so that $T(x) = P_1(x)(\mu_{12}(x) + \mu_{15}(x))$. The formulation proposed for this model, however, in terms of $f(x)$ and $T(x)$, allows for a modelling representation of the transmission of infection through sexual activity.

The complete specification of $T(x)$ is based on terms relating to the total probability of being immune and positive from AIDS at age x.

Define

$$\bar{P}_j(x) = \int_0^{x-x_0} k_j(x,z)P_j(x,z)\,dz \quad \text{for } j = 2,5 \qquad (7.10)$$

and

$$A_j(x) = \int_0^{x-x_0} \mu_{1j}(x,z)P_j(x,z)\,dz \quad \text{for } j = 2,5. \qquad (7.11)$$

Then the evaluation of $T(x)$ used in the model is

$$T(x) = P_1(x)\left(\frac{A_2(x) + A_5(x)}{P_1(x) + \bar{P}_2(x) + \bar{P}_5(x)}\right). \qquad (7.12)$$

The so-called intensities of infectivity $\mu_{12}(x,z)$ and $\mu_{15}(x,z)$ may depend on a range of covariables, and many epidemiological models have tended to use a combination of the following two elements: the frequency of sexual contact with a new partner and the conditional probability of infection from a new partner. It is assumed in this particular model that these intensities vary with the age of the person at risk, perhaps representing varying levels of sexual activity at different ages, and also with the length of time that the partner has been infected (Wilkie, 1989).

The terms $k_j(x,z)$ are used to represent the frequency with which positive and immune individuals respectively enter the pool of persons involved in sexual activity, relative to those at risk who are assumed to enter with a unit frequency.

If there were no immunes, if the infectivity intensity were a constant μ for all ages and durations and if $k_2(x,z) = 1$, then, as noted by Wilkie (1989), the density of new infections would be

$$T(x) = P_1(x)\left(\frac{\mu P_2(x)}{P_1(x) + P_2(x)}\right) \qquad (7.13)$$

which in the absence of any other transitions would lead to the common result that the infected proportion in the population would grow as a logistic function (Anderson et al., 1986).

7.2.2 Numerical solution

The same discretisation technique as described earlier in section 4.5 has been used for obtaining numerical solutions.

As described by Wilkie (1989), given a set of initial conditions, approximations to the differential equations are used to take forward the initial conditions in steps of length h, where h is some convenient fraction of a year, for example one quarter.

For states 2, 3 and 5, where the status depends on current duration, this is also subdivided into steps of h, i.e. $(0, h)$, $(h, 2h)$, etc. The corresponding transition probabilities are then estimated step to step by a discrete approximation. The integrals that appear in equations (7.7), (7.10) and (7.11) are then approximated by summations.

The model allows a flexible representation of the transition probabilities and intensities, varying by age, calendar year and where relevant also by duration in the current state (Wilkie, 1989).

7.2.3 Bases

A wide range of assumptions have been tested: for full details readers are referred to Daykin *et al.* (1988a, 1988b, 1990) and Wilkie (1988a, 1989).

Although the facility for including an immune state exists in the model, it has not been used in any of the published bases. As noted earlier in this section, those sick with AIDS are assumed not to contribute to new infections, so that, in the absence of an immune state, infection is only possible from the positives, who are assumed to contribute to the denominator of equation (7.12) to the same extent as the at risk, i.e. $k_2(x, z)$ is taken as unity for all x and z(Wilkie, 1989).

As an illustration, Daykin *et al.* (1989) consider the effects of:

$\mu_{410}(x) = \mu_{16}(x) = \mu_{27}(x)$: current national force of mortality;

μ_{14} as a function of calendar time;

μ_{12} as a function of age, duration since becoming infected and calendar time;

μ_{23} as a function of duration in the positive state (state 2);

μ_{39} as a function of age and calendar time.

Specifically, experiments with $\mu_{23}(z)$ have involved the choices

$$\left. \begin{array}{l} \mu_{23}(z) = \mathrm{Min}[\exp(-a + bz), c] \\ ab^a z^{a-1} \end{array} \right\} \qquad (7.14)$$

respectively Gompertz and Weibull in form. We note that this intensity is synonymous with the hazard rate for the incubation period distribution.

Other critical assumptions needed in this model are the proportion of the starting population assumed to be at risk or positive and the proportion of future new entrants (at age 15) assumed to be at risk or positive.

The model assumes, through the equations describing transmission from the infected to the susceptible, homogeneous mixing as far as sexual activity is concerned. However, it should be noted that the 'clear' group could be said to have zero new sexual partners per unit time while the 'at risk' and 'positive' groups have non-zero new sexual partners. Allowing the transition intensity from 'at risk' to 'positive' (via infection) to decrease over time would permit the model to attempt to represent the rate of spread of the infection from the very promiscuous to the less promiscuous as would happen in a heterogeneous (real) population.

The model has proved to be both flexible and useful in terms of its intended purposes of assisting actuarial applications and providing projections for the total population. However, it should be pointed out that the model's failure to allow for key epidemiological features of the spread of HIV and AIDS (in particular, the heterogeneity of sexual activity in the population) means that it is extremely unlikely that projections would be exactly realized. However, it has provided a benchmark set of forecasts.

The separate treatment of age cohorts contains the major restriction that infection can only be transmitted within cohorts and not between them. It would be possible to augment the model to permit some assumption about mixing of partners between cohorts – the complications caused to the numerical solutions of the resulting equations have meant that this has not been pursued.

7.2.4 Modelling the underwriting process

As discussed by Daykin *et al.* (1988b), in order to explore the consequences of different underwriting strategies, it is necessary to take into account the effect of temporary initial selection. To do this, the multiple state model has been modified to trace through individuals starting in each particular status and to identify the proportions of each such group who have progressed to other states at subsequent times.

Assumptions are needed about the efficacy of the selection process. In the absence of the immune state, the live population is divided into four live states: 1, 2, 3, 4, in Fig. 7.2. The insurer is assumed to select new policyholders from a mix of these groups, according to the criteria adopted for underwriting. Thus, an insurer applying no underwriting selection (and not experiencing any adverse selection), would accept 100% of each category and hence its new policyholders would correspond to a cross-section of the population in terms of the prevalence of these four live states.

At the other extreme, an insurer with perfectly clairvoyant underwriters may be able to operate such an efficient underwriting process as to admit only those in the clear category. In this case, the new policyholders would consist of 100% of the clears and 0% of the other groups. This would correspond to a population without AIDS and would provide a useful standard for comparison. Intermediate positions may be chosen, by specifying that the insurer includes a proportion of those in the other states, sick with AIDS, those HIV-positive and those at risk, the proportions varying in each case between 0% and 100%. Adverse selection by those who know they are at risk (or HIV-positive) can be allowed for by taking a higher proportion in that category, even to the extent of assuming more than 100% (section 7.3). Numerical illustrations are provided by Daykin *et al.* (1988b).

7.2.5 References

For more information on the model and its application and use, readers are referred to Daykin *et al.* (1988a, 1988b, 1989, 1990) and Wilkie (1988a, 1989).

7.3 DUTCH AIDS MODEL

Alting von Geusau (1990) presents a model for assessing the actuarial consequences of the AIDS risk in the Netherlands.

The model is similar in structure to that proposed by the Institute of Actuaries AIDS Working Party (and discussed in section 7.2) and is depicted in Fig. 7.3.

Differences of details include:

- the possibility of transition from 'Clear' to 'At Risk';
- the splitting of states (as described in sections 1.7 and 2.3) for those infected by HIV and sick with AIDS: the explicit separation out of successive durations of occupancy.

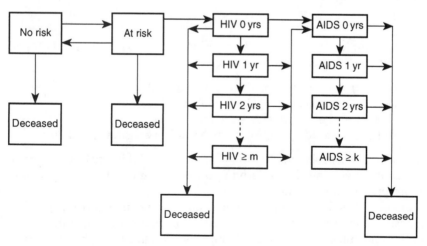

Figure 7.3 The general AIDS model: Alting von Geusau (1989).

For practical purposes, it is assumed that there is a maximum duration for identifying the substrates. For HIV, this assumption is that the maximum is eight years, so that durations above eight years are not distinguished from eight years. For those sick with AIDS, the maximum is five years.

Alting von Geusau uses a discrete time formulation as in Chapter 2. We define

m — the number of states considered;

$\mathbf{n}(x, t)$ — the vector with the number of individuals age x at time t who are in one of the m states;

$\mathbf{P}(x, t)$ — the $m \times m$ matrix with transitions probabilities for an individual at age x at time t where the (i, j) element represents the probability of transition from state i at the beginning of the year to state j at the end.

Then, the vector $\mathbf{n}$ satisfies the recursive relationship

$$\mathbf{n}'(x + 1, t + 1) = \mathbf{n}'(x, t)\mathbf{P}(x, t). \tag{7.15}$$

From the $\mathbf{n}(x, t)$ vectors the $\mathbf{N}(t)$ matrix can be constructed, with the (i, j) entry representing the number of persons at time t at age i in state j. The population at time t is then the result of the initial population at time t_0 represented by the matrix $\mathbf{N}(t_0)$ and the choice of parameters determining the transition probabilities in the matrix $\mathbf{P}$ and assumptions about the flow of new entrants.

As in section 7.2.4, it is possible to consider the effect on an insured population by introducing a vector $\mathbf{w}(t)$, with the ith entry equal to the propensity to insure in state i. Then the relevant insured population is given by the vector

$$\mathbf{z}(t) = \mathbf{N}(t) \cdot \mathbf{w}(t). \tag{7.16}$$

As an example, if we consider the four live states: no risk, at risk, HIV and sick with AIDS, then possible choices of w would be

(1 0 0 0)   protection against almost all risks;
(1 1 0 0)   protection against 'bad' risks;
(1 2 4 10)   heavy adverse selection;
(1 1 1 1)   solidarity through a compulsory insurance scheme.

The transition probabilities satisfy a series of differential and integral equations involving the underlying transitions intensities, analogous to those presented in section 7.2.1.

As in section 7.2.2, the probabilities are then calculated from specified transition intensities by using numerical procedures to solve the differential and integral equations. Alting von Geusau bases his numerical procedures on Simpson's rule for approximate integration.

7.4 WALTER REED STAGING METHOD (WRSM) MODEL

If we consider a life who is HIV positive but is otherwise healthy and asymptomatic, then a method commonly used to describe such a life's possible progression to AIDS is the Walter Reed Staging Method (WRSM), described in detail by Redfield *et al.* (1986).

The WRSM classifies patients who have tested HIV positive into four stages along the route to full-blown AIDS, rather than grouping patients according to their complications. For completeness, we have added two more stages, 'at risk' and 'death'. These stages (labelled 1–6) are described below (as in Panjer, 1988):

Stage 1 (At risk): healthy persons at risk for HIV positive infection, but testing negative;

Stage 2 (HIV positive): otherwise asymptomatic persons testing HIV positive;

Stage 3 (LAS): persons with HIV infection and lymphademopathy syndrome (LAS), together with moderate cellular immune deficiency;

Stage 4 (ARC): patients with HIV infection and LAS, plus severe cellular immune deficiency (AIDS-Related Complex, or ARC);

Stage 5 (AIDS): patients with AIDS;

Stage 6 (Death): patients who died 'of AIDS'.

A number of actuarial models of AIDS have been developed, most notably in the US using this classification as a principal building block.

In the approach of Cowell and Hoskins (1988) and Panjer (1988), it is assumed that individuals progress through the successive stages in order (i.e. sequentially) and do not return to a previous stage, i.e. a person's medical condition can remain the same or deteriorate but not improve. It is also assumed that death prior to stage 5 (i.e. full blown AIDS) is not permitted. This model is depicted in Fig. 7.4, with the states corresponding to the WRSM stages. Panjer uses a rigorous statistical approach to the model specification and estimation of parameters (using maximum likelihood methods to estimate the transitions intensities from data provided by the Frankfurt study: Brodt *et al.* (1986)). Panjer uses a continuous time Markov process (as in the section 7.2),

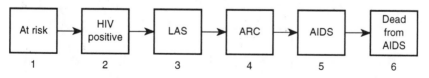

Figure 7.4 Cowell and Hoskins model and Panjer model, based on Walter Reed Staging Method (WRSM).

assuming that the transition intensities for a person in a given stage of HIV infection depend only on the current state of occupancy, and not upon other factors such as age, sex and the duration of time spent in that state. This assumption simplifies considerably the mathematical presentation and use is made of the property that the random variables representing the time spent in state j are then stochastically independent with exponential distributions. Further, the 'memoryless' property of the exponential distribution means that the length of time that a person has been in the current state is not relevant and the expected time of progression to the next state is the same for all persons in that state (i.e. it is independent of the time already in the current state; Panjer (1988)).

This representation also avoids the technical difficulty of modelling the infection mechanism (as noted in sections 7.2 and 7.3).

The model has been extended by Ramsay (1989) who incorporates a further state to accommodate mortality from states 1–4 and from state 5 due to a cause of death unrelated to AIDS. This is depicted in Fig. 7.5. For transitions into state 7 (from state j, for $j = 1, \ldots, 5$), it is assumed that the transition intensity depends only on the state of occupancy. As above, the 'memoryless' property means that we can speak in terms of the future time spent in a state without having to condition on the amount of time already spent in that state.

Ramsay uses this formulation to develop expressions for the traditional life insurance functions, for example, net single premiums, actuarial present values of annuities, reserves and the distribution of the loss random variables, for a life in state i at the time of issue of the policy. Ramsay's approach makes use of either the backward Kolmogorov differential equations or a consideration of the moments of the random variables $T_i^{(d)}$, the future lifetime until death (from any cause) for a life currently in state i.

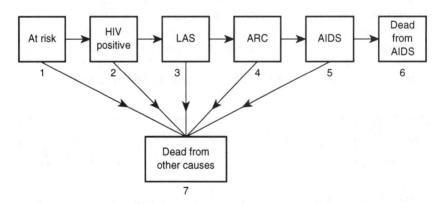

Figure 7.5 Ramsay model based on WRSM.

7.5 HABERMAN MODEL AND DISABILITY INSURANCE

The Institute of Actuaries AIDS Working Party model described in section 7.2 is complex and hence it is difficult to discern the important properties and identify the key parameters. A modified version, along the lines of section 7.4, has been proposed with simplifications (see Haberman, 1992, 1995) so that Markov processes techniques can be applied: see Fig. 7.6.

As in section 7.4, the model is a continuous time Markov process with constant transitions intensities μ_{ij} for transition from state i to state j for the permissible values of i and j in Fig. 7.6. Also, as before, the important memoryless property applies here.

We now consider how the model of the Institute of Actuaries AIDS Working Party (as described in section 7.2) may be modified to fit in with these assumptions.

1. The immune state is removed.
2. The viewpoint is changed from that of the population as a whole to that of an individual male at risk who is considered to progress from state to state over time. We are then concerned not with the spread of HIV in a population, but with the outcome for a particular individual.
3. It is assumed (as in section 7.4) that all transition intensities are constants, independent of attained age, duration in current state, and secular time. We acknowledge that this assumption contradicts the arguments of section 7.2 (and Daykin *et al.*, 1988b, 1990) that emphasize the importance of these variables, in particular attained age, to an actuarial assessment of the effects of HIV and AIDS on survival prospects. Two arguments support this seemingly extreme assumption.
   (a) The magnitude of the AIDS-related transition intensities outweighs the normal age-related mortality risk. Indeed, many of

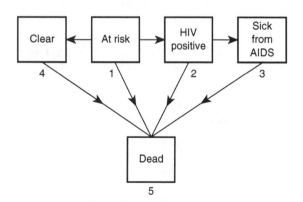

Figure 7.6 Modified AIDS Working Party model (Version 1).

the AIDS Working Party simulations assume intensities that do not vary with respect to age.

(b) The desire to reach some analytical results may require, at least initially, some heroic assumptions. We believe that the results may be, nevertheless, of value in pricing and reserving for disability insurance.

4. It is assumed (as in section 7.4) that the transition intensity from the at-risk state to the seropositive state is constant and does not depend on the numbers of persons infected. This again is a simplifying assumption to keep the resulting mathematical manipulations tractable. As noted by Daykin *et al.* (1990), a constant transition intensity from at-risk to seropositive would be consistent with the exponential development of new cases of AIDS in the early stages of the epidemic.

Remark
To allow for the effect of heterogeneity of risk and behavioural change, it would be reasonable to postulate an intensity that decreases with time as the epidemic develops. This assumption is not pursued here on the grounds of mathematical tractability.

An advantage to making these simplifying assumptions to the original model is that the model is now flexible enough to accommodate the heterosexual transmission of HIV and the development of AIDS without the restriction to the male homosexual population mentioned in section 7.2.

Given these assumptions, the next step is to determine the transition probabilities. To this end, we let $P_{ij}(t)$ be the transition probability that a life now in state i will be in state j at t years from now. There are a number of different ways to set up the equations for the required transition probabilities: as in equation (1.20) we use the Kolmogorov backward system of differential equations. Because of the assumption of constant transition intensities, we obtain simple recursive solutions to these equations.

We consider an insurance policy issued to a life in state i at the time of issue, i.e., at $t = 0$. It can then be proved that

$$P_{11}(t) = e^{-\mu_1 t} \tag{7.17}$$

$$P_{22}(t) = e^{-\mu_2 t} \tag{7.18}$$

$$P_{33}(t) = e^{-\mu_{35} t} \tag{7.19}$$

$$P_{44}(t) = e^{-\mu_{45} t} \tag{7.20}$$

$$P_{12}(t) = \frac{\mu_{12}}{\mu_1 - \mu_2}(e^{-\mu_2 t} - e^{-\mu_1 t}) \tag{7.21}$$

$$P_{23}(t) = \frac{\mu_{23}}{\mu_2 - \mu_{35}} (e^{-\mu_{35}t} - e^{-\mu_2 t}) \tag{7.22}$$

$$P_{13}(t) = \frac{\mu_{12}\mu_{23}}{(\mu_1 - \mu_2)(\mu_1 - \mu_{35})(\mu_2 - \mu_{35})}$$
$$\times [(\mu_2 - \mu_{35}) e^{-\mu_1 t} - (\mu_1 - \mu_{35}) e^{-\mu_2 t} + (\mu_1 - \mu_2) e^{-\mu_{35}t}] \tag{7.23}$$

$$P_{14}(t) = \frac{\mu_{14}}{\mu_1 - \mu_{45}} (e^{-\mu_{45}t} - e^{-\mu_1 t}) \tag{7.24}$$

where $\mu_1 = \mu_{12} + \mu_{14} + \mu_{15}$ and $\mu_2 = \mu_{23} + \mu_{25}$ (as in section 1.4). We note from the model structure that $P_{ij}(t) = 0$ for $i > j$ and that $P_{24}(t) = P_{34}(t) = 0$.

The associated probabilities of dying (being in state 5) are given by the following expressions:

$$P_{15}(t) = 1 - \sum_{j=1}^{4} P_{1j}(t)$$

$$P_{25}(t) = 1 - \sum_{j=2}^{3} P_{2j}(t)$$

$$P_{35}(t) = 1 - e^{-\mu_{35}t}$$

$$P_{45}(t) = 1 - e^{-\mu_{45}t}.$$

7.5.1 Application to disability insurance

We consider the application to disability insurance policies as in Chapter 3 which have a deferred period of f in their design.

Initially, we assume that a sickness claim is admitted only when full AIDS develops, i.e., the policyholder is in state 3. To calculate actuarial values, we then need the probability that an individual starting in state 1 is sick throughout the time interval $(t - f, t)$, i.e. that the underlying stochastic process is in state 3 throughout this time interval. In the absence of the deferred period, the probability that the policy-holder is in state 3 at time t given that the policyholder was in state 1 at time 0 is $P_{13}(t)$. It can be verified, by appealing to the Markov property, that $P_{13}(t)$ can be written as

$$P_{13}(t) = \int_0^t P_{12}(u)P_{33}(t - u)\mu_{23} \, du. \tag{7.25}$$

We can adapt equation (7.25) to allow for the presence of a deferred period. We define $q_f(t)$ to be the probability that a person in state 1 at time zero is in state 3 throughout the time interval $[t - f, t]$. Adapting the above integral definition, we can write the following:

$$q_f(t) = \int_0^{t-f} P_{12}(u)P_{33}(t-u)\mu_{23} \, du \quad \text{for } t > f, \qquad (7.26)$$

which corresponds to equation (3.11).

To deal with sickness claims that are paid while the policyholder is in state 2 or state 3, we define $r_f(t)$ to be the probability that a person in state 1 at time zero is in state 2 or state 3 throughout the time interval $[t-f, t]$. Then

$$r_f(t) = \int_0^{t-f} P_{11}(u)(P_{22}(t-u) + P_{23}(t-u))\mu_{12} \, du \quad \text{for } t > f. \qquad (7.27)$$

Given the earlier results (equations (7.19) and (7.21)), it follows that equation (7.26) becomes:

$$q_f(t) = \frac{\mu_{12}\mu_{23}\, e^{-\mu_{35}t}}{(\mu_1 - \mu_2)} \left[\frac{1 - e^{-(\mu_2 - \mu_{35})(t-f)}}{(\mu_2 - \mu_{35})} - \frac{1 - e^{-(\mu_1 - \mu_{35})(t-f)}}{(\mu_1 - \mu_{35})} \right]. \qquad (7.28)$$

Similarly, from equations (7.17), (7.18) and (7.22), equation (7.27) becomes:

$$r_f(t) = \frac{\mu_{12}}{(\mu_2 - \mu_{35})} \left[\frac{\mu_{23}\, e^{-\mu_{35}t}(1 - e^{-(\mu_1 - \mu_{35})(t-f)})}{(\mu_1 - \mu_{35})} \right. $$
$$\left. + \frac{(\mu_{23} - \mu_{35})\, e^{-\mu_2 t}(1 - e^{-(\mu_1 - \mu_2)(t-f)})}{(\mu_1 - \mu_2)} \right]. $$

7.5.2 Valuation functions

Following Daykin *et al.* (1988a), we recognize that a major difficulty in estimating the impact of HIV infection and AIDS on disability insurance business is knowing at what stage a claim will be presented to the insurer. For the purposes of illustration of the methodology and the results, we consider here two extreme cases.

Case 1

We assume that a claim is admitted only when full AIDS develops. In the case of a disability insurance policy with a deferred period equal to f, we assume that no benefit is payable until a time f after AIDS has developed (i.e., after entry to state 3).

Case 2

We make the equally extreme assumption that claims are admitted on the basis of HIV seropositivity alone (i.e., on entry to state 2) without requiring evidence of AIDS or any of the intermediate stages.

Let $A_j(n,f)$ be the actuarial present value (under Case j) of a disability income benefit of one unit (per year) in an n year policy (with the n years measured from the inception of the policy) with deferred period of f, for $j = 1, 2$. It follows that:

$$A_1(n,f) = \int_f^n e^{-\delta t} q_f(t)\, dt \quad \text{Case 1;}$$

$$= \mu_{12}\mu_{23}[(a_1 + a_2 - a_3)\, e^{-(\delta + \mu_{35})f} - a_1\, e^{-(\delta + \mu_{35})n}$$

$$- a_2\, e^{-(\mu_1 - \mu_{35})f}\, e^{-(\delta + \mu_1)n} + a_3\, e^{(\mu_2 - \mu_{35})f}\, e^{-(\delta + \mu_2)n}] \tag{7.29}$$

$$A_2(n,f) = \int_f^n e^{-\delta t} r_f(t)\, dt \quad \text{Case 2;}$$

$$= \mu_{12}[\mu_{23}a_1 (e^{-(\delta + \mu_{35})f} - e^{-(\delta + \mu_{35})n})$$

$$+ a_3(\mu_{25} - \mu_{35})(e^{-(\delta + \mu_2)f} - e^{-(\delta + \mu_2)n})$$

$$- (\mu_{23}a_4\, e^{(\mu_1 - \mu_{35})f} + a_5(\mu_{25} - \mu_{35})\, e^{(\mu_1 - \mu_2)f})$$

$$\times (e^{-(\delta + \mu_1)f} - e^{-(\delta + \mu_1)n})] \tag{7.30}$$

where the a_ks are constants with their reciprocals given by

$$a_1^{-1} = (\mu_1 - \mu_{35})(\mu_2 - \mu_{35})(\delta + \mu_{35})$$

$$a_2^{-1} = (\mu_1 - \mu_{35})(\mu_1 - \mu_2)(\delta + \mu_1)$$

$$a_3^{-1} = (\mu_2 - \mu_{35})(\mu_1 - \mu_2)(\delta + \mu_2)$$

$$a_4^{-1} = (\mu_2 - \mu_{35})(\mu_1 - \mu_{35})(\delta + \mu_1)$$

$$a_5^{-1} = (\mu_2 - \mu_{35})(\mu_1 - \mu_2)(\delta + \mu_1).$$

From these results, it would be possible to investigate the explicit forms for the partial derivatives of $A_1(n,f)$ and $A_2(n,f)$ with respect to any of the underlying parameters. To illustrate, we present some numerical values of $A_1(n,f)$ and $A_2(n,f)$ based on equations (7.29) and (7.30) for different combinations of some of the underlying parameters.

Following Daykin *et al.* (1990), we set $\mu_{35} = 0.35$ and $\mu_{15} = \mu_{45} = 0.001$ (corresponding approximately to the force of mortality for a male age 30–34 according to English Life Table No. 14, based on decennial data for 1980–82). We also set $\mu_{14} = 0.10$ and $\delta = 0.07$. Tables 7.1 to 7.3 present the magnitudes of A_1 and A_2 for the values of $\mu_{12}, \mu_{23}, \mu_{25}, n$ and f shown. For convenience, $\mu_{12} = \mu_{23}$ in this presentation. The results indicate that A_1 and A_2 both increase with increasing n, decreasing f, decreasing μ_{25} and increasing $\mu_{12} (= \mu_{23})$. They further indicate the relative sensitivities of A_1 and A_2 to changes in these parameters and also demonstrate that the ratio A_1/A_2 decreases as $\mu_{12} = \mu_{23}$ increases. These results are as expected.

Table 7.1 Present value of disability insurance benefits: $\mu_{12} = \mu_{23} = 0.001$

Deferred period: 3 months Policy terms (years)	5	10	15	20
$\mu_{25} = 0.001$				
100 A_1	0.00077	0.00341	0.00655	0.00935
100 A_2	0.764	2.242	3.673	4.842
Ratio	992	716	561	518
$\mu_{25} = 0.01$				
100 A_1	0.00076	0.00332	0.00631	0.00889
100 A_2	0.752	2.172	3.511	4.569
Ratio	989	654	566	514
$\mu_{25} = 0.05$				
100 A_1	0.00073	0.00299	0.00538	0.00721
100 A_2	0.700	1.900	2.904	3.599
Ratio	958	635	540	499

Deferred period: 6 months Policy terms (years)	5	10	15	20
$\mu_{25} = 0.001$				
100 A_1	0.00061	0.00293	0.00576	0.00830
100 A_2	0.687	2.127	3.543	4.705
Ratio	1130	726	615	567
$\mu_{25} = 0.01$				
100 A_1	0.00061	0.00286	0.00555	0.00790
100 A_2	0.675	2.059	3.381	4.432
Ratio	1110	720	609	561
$\mu_{25} = 0.05$				
100 A_1	0.00058	0.00258	0.00474	0.00642
100 A_2	0.623	1.788	2.776	3.465
Ratio	1070	693	586	540

We can compare these results with those given by Daykin *et al.* (1988a) for the actuarial value of the additional sickness benefits under a PHI policy allowing for the two extreme cases described above. Daykin *et al.* (1988a) use different morbidity and mortality assumptions (intermediate between the sets underlying Tables 7.2 and 7.3). It is impossible to rerun their full model on modified assumptions; however, we can consider from their appendix tables the values of $A_1(n,f)$ and $A_2(n,f)$ and the ratio of the present values under Cases 1 and 2 for comparison with Tables 7.1 to 7.3. The details appear in Table 7.4 for a deferred period of six months and two alternative terminal ages. The magnitude of $A_1(n, f)$ and $A_2(n, f)$ and the ratios are intermediate between those appearing in Tables 7.2 and 7.3 and display similar trends. In particular, we note the stability of the ratios as we consider different age ranges.

Table 7.2 Present value of disability insurance benefits: $\mu_{12} = \mu_{23} = 0.01$

Deferred period: 3 months				
Policy terms (years)	5	10	15	20
$\mu_{25} = 0.001$				
$100 \, A_1$	0.0755	0.3257	0.6130	0.8586
$100 \, A_2$	7.507	21.567	34.655	44.890
Ratio	99.5	66.2	56.5	52.3
$\mu_{25} = 0.01$				
$100 \, A_1$	0.0746	0.3178	0.5906	0.8168
$100 \, A_2$	7.385	20.911	33.149	42.418
Ratio	99.0	65.8	56.1	51.9
$\mu_{25} = 0.05$				
$100 \, A_1$	0.0710	0.2863	0.5049	0.6660
$100 \, A_2$	6.874	18.320	27.515	33.621
Ratio	96.9	64.0	54.4	50.5

Deferred period: 6 months				
Policy terms (years)	5	10	15	20
$\mu_{25} = 0.001$				
$100 \, A_1$	0.0600	0.2802	0.5396	0.7629
$100 \, A_2$	6.745	20.458	33.405	43.584
Ratio	112	73.0	61.9	57.1
$\mu_{25} = 0.01$				
$100 \, A_1$	0.0593	0.2736	0.5201	0.7265
$100 \, A_2$	6.625	19.806	31.904	41.117
Ratio	112	72.3	61.3	56.6
$\mu_{25} = 0.05$				
$100 \, A_1$	0.0566	0.2472	0.4458	0.5937
$100 \, A_2$	6.125	17.231	26.288	32.339
Ratio	108	70.0	59.0	54.5

Similarly, expressions for the present value of premiums and expenses may be developed, including the value of expenses related to the timing of the payment of the sickness benefit. Also, allowance may be made for a waiver of premium benefits and for stepped sickness benefits, i.e., a level of sickness income that depends on the current duration of sickness.

For example, we consider allowing for a waiver of premium benefit under Case 1, where a sickness claim is admitted only when a transition is made to state 3 (the development of full AIDS). Let AP_t be the annual premium payable at time $t = 0, 1, \ldots, n - 1$. Then the actuarial value of the annual premiums ($APVP$) is given by:

$$APVP = \sum_{t=0}^{n-1} e^{-\delta t} AP_t \left(\sum_{j=1}^{4} P_{1j}(t) - q_f(t) \right).$$

Table 7.3 Present value of disability insurance benefits: $\mu_{12} = \mu_{23} = 0.10$

Deferred period: 3 months Policy terms (years)	5	10	15	20
$\mu_{25} = 0.001$				
100 A_1	6.092	21.257	33.260	40.081
100 A_2	63.091	150.480	206.611	233.975
Ratio	10.4	7.18	6.18	5.84
$\mu_{25} = 0.01$				
100 A_1	6.026	20.793	32.223	38.545
100 A_2	62.129	146.437	198.228	224.077
Ratio	10.3	7.04	6.15	5.81
$\mu_{25} = 0.05$				
100 A_1	5.745	18.916	28.203	32.804
100 A_2	58.097	130.326	170.037	187.542
Ratio	10.1	6.88	6.03	5.72

Deferred period: 6 months Policy terms (years)	5	10	15	20
$\mu_{25} = 0.001$				
100 A_1	4.895	18.475	29.524	35.867
100 A_2	56.689	142.179	196.820	225.057
Ratio	11.6	7.70	6.67	6.27
$\mu_{25} = 0.01$				
100 A_1	4.845	18.081	28.616	34.503
100 A_2	55.748	138.163	189.466	215.189
Ratio	11.5	7.64	6.62	6.24
$\mu_{25} = 0.05$				
100 A_1	4.630	16.485	25.088	29.398
100 A_2	51.809	122.173	161.403	178.784
Ratio	11.2	7.41	6.43	6.08

If the deferred period were f_1 and the level of sickness benefit were B per annum for sickness of durations u where $f_1 < u \le f_1 + f_2$ and the level of sickness benefit were C per annum ($< B$) for sickness of durations u where $u > f_1 + f_2$, then the actuarial value of the benefits ($APVB$) under Case i (where $i = 1, 2$) would be given by

$$APVB_i = BA_i(n, f_1) + (C - B)A_i(n, f_1 + f_2). \tag{7.31}$$

7.5.3 Emerging costs

Although the model is of a time-continuous type, it can be used for valuations in a time-discrete context, as in the definition of APVP (see section 2.1.1) and, as in section 2.5, we can also consider emerging costs and cash flow calculations. For illustration, we choose a simplified example. We consider a non-profit disability insurance policy for a term

Table 7.4 Present values of disability insurance benefits according to the Daykin *et al.* (1988a) model; deferred period six months

Terminating age of policy: 60 Policy terms (years)	10	15	20
Assumption A			
100 A_1	0.067	1.76	2.18
100 A_2	16.77	24.26	26.12
Approximate ratio	25	14	12
Assumptions BC			
100 A_1	0.48	1.02	1.28
100 A_2	9.49	13.88	15.25
Approximate ratio	20	14	12
Assumption F			
100 A_1	0.65	1.36	1.71
100 A_2	6.88	9.33	10.11
Approximate ratio	11	7	6

Terminating age of policy: 65 Policy terms (years)	10	15	20
Assumption A			
100 A_1	0.34	1.57	2.13
100 A_2	8.81	22.36	25.70
Approximate ratio	26	14	12
Assumptions BC			
100 A_1	0.25	0.90	1.25
100 A_2	4.71	12.13	14.83
Approximate ratio	19	13	12
Assumption F			
100 A_1	0.38	1.25	1.65
100 A_2	3.96	8.54	9.93
Approximate ratio	10	7	6

Note on assumptions made by Daykin *et al.* (1998a):

A	$\mu_{13}, \mu_{23}, \mu_{45}$	England and Wales population mortality
	μ_{12}	0.7 at ages 25–50, reducing to zero at ages 15 and 70
	μ_{23}	Max $[\exp(-8.4 + 14t), 2.5]$ where t = duration in state 2
	μ_{25}	Normal mortality $+0.7$
	μ_{14}	0
BC	$\mu_{13}, \mu_{23}, \mu_{45}$	As for projection A
	μ_{12}	As for projection A, but reducing linearly from 1987 to 1992 to half initial intensity at all ages
	μ_{23}	As for projection A
	μ_{25}	As for projection A
	μ_{14}	0.10
F	$\mu_{13}, \mu_{23}, \mu_{45}$	As for projections A and BC
	μ_{12}	As for projection BC
	μ_{23}	As for projections A and BC
	μ_{25}	Normal mortality $+0.35$
	μ_{14}	As for projection BC.

of n years with a zero deferred period (to simplify the algebra). The policy has annual premiums, AP_t, being paid at times $t = 0, 1, \ldots, n - 1$. The premium is not paid if the policyholder is sick at time t. We use Case 1 for the definition of sickness for purposes of illustration.

We take the benefits provided by the policy to be:

- a death benefit of D_t payable at the end of the policy year if the policyholder dies during the tth policy year (for $t = 1, 2, \ldots, n$);
- an income benefit of B_t payable at the end of the tth year if the policyholder is then alive and sick (for $t = 1, 2, \ldots, n$).

We assume that the rate of interest in the tth policy year is i_t and that the expected cash flow for the tth policy year per policy alive and in state j at the start of the tth policy year is $CF_t^{(j)}$. It follows that

$$
CF_t^{(j)} = \begin{cases} (1+i_t)AP_{t-1} - P_{j3}(1)B_t - P_{j5}(1)D_t & \text{if } j = 1 \text{ or } 2 \\ \qquad\qquad - P_{33}(1)B_t - P_{35}(1)D_t & \text{if } j = 3 \\ (1+i_t)AP_{t-1} \qquad\qquad - P_{45}(1)D_t & \text{if } j = 4. \end{cases}
$$

Then, we define the expected costs (or cash flow) for the tth policy year per policy originally issued to be EC_t where

$$
EC_t = \sum_{j=1}^{4} CF_t^{(j)} P_{1j}(t - 1).
$$

The above development can be extended to allow for deferred periods, varying benefits, and transition intensities that are functions of attained age. If the AP_ts were net premiums (in the traditional sense of the term), and the i_t constants, then the equation of value for such a policy would, by definition, be given by

$$
\sum_{t=1}^{\infty} e^{-\delta t} EC_t = 0.
$$

This provides an extension to the results given by Hare and McCutcheon (1991) in respect of conventional life insurance profit testing.

7.5.7 Further modifications to the model

Separating incidence of disability

The discussion earlier is based on the model depicted in Fig. 7.6 and considers the two extreme cases (Case 1 and Case 2) of timing of a claim for disability income. A more satisfactory approach would be to recognize explicitly the existence of an intermediate state between HIV positive and AIDS for those who are sick and, hence, eligible to claim. Figure 7.7 depicts the new model needed, with the states renumbered. A disability income claim would be accepted once a policyholder has

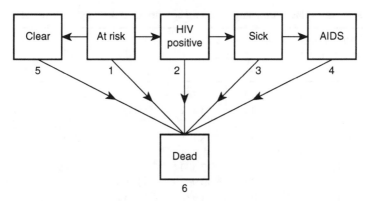

Figure 7.7 Modified AIDS Working Party model (Version 2).

entered state 3 (and the income benefit would be payable while he/she occupies either states 3 or 4).

This modified model leads to no conceptual difficulties. We still must develop the Kolmogorov system of differential equations and solve for the transition probabilities $P_{ij}(t)$. The solutions are similar to those given in equations (7.17) to (7.24); see Ramsay (1989). For example,

$$P_{33}(t) = e^{-\mu_3 t} \tag{7.32}$$

$$P_{34}(t) = \frac{\mu_{34}}{\mu_3 - \mu_{46}}(e^{-\mu_{46} t} - e^{-\mu_3 t}) \tag{7.33}$$

where $\mu_3 = \mu_{34} + \mu_{36}$.

To deal with sickness claims that are paid while the policyholder is in state 3 or state 4 in the presence of a deferred period f, we define $s_f(t)$ to be the probability that a person in state 1 at time 0 is in state 3 or state 4 throughout the time interval $[t-f, t]$ for $t > f$. Then, as for equations (7.26) and (7.27), we obtain

$$s_f(t) = \int_0^{t-f} P_{12}(u)(P_{33}(t-u) + P_{34}(t-u))\mu_{23}\,du \quad \text{for } t > f. \tag{7.34}$$

Substitution from equations (7.21), (7.32) and (7.33) leads to the following:

$$\begin{aligned}
s_f(t) = {}& \frac{\mu_{12}\mu_{23}\,e^{-\mu_3 t}}{(\mu_1 - \mu_2)}\left[\frac{(1 - e^{-(\mu_2 - \mu_3)(t-f)})}{(\mu_2 - \mu_3)} - \frac{(1 - e^{-(\mu_1 - \mu_3)(t-f)})}{(\mu_1 - \mu_3)}\right] \\
&+ \frac{\mu_{12}\mu_{23}\mu_{34}\,e^{-\mu_{46} t}}{(\mu_1 - \mu_2)(\mu_3 - \mu_{46})}\left[\frac{(1 - e^{-(\mu_2 - \mu_{46})(t-f)})}{(\mu_2 - \mu_{46})} - \frac{(1 - e^{-(\mu_1 - \mu_{46})(t-f)})}{(\mu_1 - \mu_{46})}\right] \\
&- \frac{\mu_{12}\mu_{23}\mu_{34}\,e^{-\mu_3 t}}{(\mu_1 - \mu_2)(\mu_3 - \mu_{46})}\left[\frac{(1 - e^{-(\mu_2 - \mu_3)(t-f)})}{(\mu_2 - \mu_3)} - \frac{(1 - e^{-(\mu_1 - \mu_3)(t-f)})}{(\mu_1 - \mu_3)}\right].
\end{aligned}$$

$$\tag{7.35}$$

Table 7.5 Present value of disability insurance benefits: modified model. Values of 100 A_3

μ_{26}	μ_{36}	μ_{23}	μ_{34}	5	10	15	20
Deferred period: three months							
Policy terms (years)				5	10	15	20
0.01	0.05	0.1	0.1	8.702	37.190	66.396	86.776
0.01	0.05	0.1	0.2	8.454	34.488	58.916	74.394
0.01	0.05	0.2	0.1	15.577	60.273	99.365	122.436
0.01	0.05	0.2	0.2	15.121	55.690	87.587	104.136
0.05	0.10	0.1	0.1	7.784	30.408	50.480	62.490
0.05	0.10	0.1	0.2	7.600	28.602	45.928	55.514
0.05	0.10	0.2	0.1	13.962	49.696	76.950	90.890
0.05	0.10	0.2	0.2	13.622	46.619	69.704	80.376
Deferred period: six months							
Policy terms (years)				5	10	15	20
0.01	0.05	0.1	0.1	7.384	34.163	62.369	82.265
0.01	0.05	0.1	0.2	7.139	31.467	54.898	69.892
0.01	0.05	0.2	0.1	13.289	55.606	93.641	116.331
0.01	0.05	0.2	0.2	12.837	51.032	81.874	98.043
0.05	0.10	0.1	0.1	6.563	27.743	47.066	58.757
0.05	0.10	0.1	0.2	6.380	25.942	42.519	51.788
0.05	0.10	0.2	0.1	11.833	45.521	71.942	85.607
0.05	0.10	0.2	0.2	11.496	42.450	64.705	75.102

The actuarial value of a disability income benefit of one unit (per year) on an n year policy with deferred period of f would be

$$A_3(n, f) = \int_f^n e^{-\delta t} s_f(t) \, dt,$$

which is evaluted in the Appendix.

As an illustration of the numerical effect of separating the incidence of disability and receipt of the income benefit from the onset of AIDS, we present some sample values of $A_3(n, f)$ in Table 7.5.

We retain the parameter values used in earlier tables ($\mu_{16} = \mu_{56} = 0.001$, $\mu_{15} = 0.10$, $\delta = 0.07$, $\mu_{46} = 0.35$); we focus on $\mu_{12} = 0.10$ for direct comparison with Tables 7.3 and 7.4. Values have been chosen such that $\mu_i \neq \mu_j$ for $i \neq j$. As expected, the values of $A_3(n, f)$ are intermediate between the two extreme estimates of $A_1(n, f)$ and $A_2(n, f)$ presented earlier. We note the extent to which $A_3(n, f)$ increases with increasing n and decreases with increasing f, decreases with increasing forces of mortality, and its relative sensitivity to the choice of μ_{23} and relative insensitivity to the choice of μ_{34}. As expected, $A_3(n, f)$ increases with increasing μ_{23} (representing the rate of flow into the claiming state) and decreases with increasing μ_{34}. Also as expected, $A_3(n, f)$ increases with increasing μ_{34} (representing part of the rate of flow out of the claiming state). Space constraints prevent our analysing further the sensitivities of $A_3(n, f)$.

Dependence on time of occupancy

It would be more realistic to allow some transition intensities to depend on the time spent in the current state since the latest transition into that state. This idea of duration dependence leads to the introduction of semi-Markov processes, as discussed in section 1.6.

The semi-Markov process can be described by a pair of continuous time stochastic processes $\{S(x), Z(x)\}$ for $x \geq 0$. Let $S(x)$ represent the state of an individual at time (or age) x where $S(x) \in 1, 2, \ldots, N$. Let $Z(x)$ denote the duration for an individual at time x of the temporary stay so far in the current state, i.e.

$$Z(x) = \max\{z : z \leq x \text{ and } S(x - u) = S(x) \text{ for all } u \in [0, z]\}.$$

The event $\{S(x) = j \text{ and } Z(x) = z\}$ represents an individual being in state j at time x with a duration of z since the last transition into state j.

We can then define transition intensities and probabilities and construct equations for the latter (which will be mixed integro-differential equations). As noted in section 1.6, the resulting expressions are complex. A useful approximation to the semi-Markov model is to introduce a number of substates. This idea has been introduced in section 1.7 and has been applied to actuarial problems by Norberg (1988) in considering select survival models and by Amsler (1968) and Jones (1994) in considering multiple state models (see also section 1.7.3, for an example).

The replacement of a state by a pair of states labelled 'stable' and 'unstable' together with the transition intensities that are independent of the time spent within each substate mimics approximately the behaviour of a semi-Markov model. Figure 7.8 illustrates part of such a model structure where we seek to approximate the duration dependence for the transition from state 1 to state 2. State 1 becomes 'HIV positive: unstable', and we add a further state 7, 'HIV positive: stable'. This modification can be implemented for the model discussed earlier and presented in Figs 7.6 and 7.7.

7.5.8 Further comments

Further discussion of the disability insurance implications of the model is provided by Habermann (1995). A corresponding discussion of life insurance implications of this model, parallel to that of Ramsay (1989), is provided by Haberman (1992).

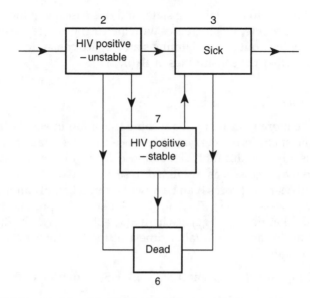

Figure 7.8 Extract from Modified AIDS Working Party model (Version 3).

APPENDIX

It is straightforward to show that

$$
\begin{aligned}
A_3(n,f) = {} & \frac{\mu_{12}\mu_{23}(\mu_{36} - \mu_{46})(e^{-(\delta+\mu_3)f} - e^{-(\delta+\mu_3)n})}{(\mu_3 - \mu_{46})(\mu_2 - \mu_3)(\mu_1 - \mu_3)(\delta + \mu_3)} \\[2mm]
& + \frac{\mu_{12}\mu_{23}\mu_{34}(e^{-(\delta+\mu_{46})f} - e^{-(\delta+\mu_{46})n})}{(\mu_3 - \mu_{46})(\mu_2 - \mu_{46})(\mu_1 - \mu_{46})(\delta + \mu_{46})} \\[2mm]
& - \frac{\mu_{12}\mu_{23}(\mu_{36} - \mu_{46})\,e^{(\mu_2-\mu_3)f}(e^{-(\delta+\mu_2)f} - e^{-(\delta+\mu_2)n})}{(\mu_1 - \mu_2)(\mu_2 - \mu_3)(\mu_3 - \mu_{46})(\delta + \mu_2)} \\[2mm]
& + \frac{\mu_{12}\mu_{23}(\mu_{36} - \mu_{46})\,e^{(\mu_1-\mu_3)f}(e^{-(\delta+\mu_1)f} - e^{-(\delta+\mu_1)n})}{(\mu_1 - \mu_2)(\mu_1 - \mu_3)(\mu_3 - \mu_{46})(\delta + \mu_1)} \\[2mm]
& - \frac{\mu_{12}\mu_{23}\mu_{34}\,e^{(\mu_2-\mu_{46})f}(e^{-(\delta+\mu_2)f} - e^{-(\delta+\mu_2)n})}{(\mu_1 - \mu_2)(\mu_3 - \mu_{46})(\mu_2 - \mu_{46})(\delta + \mu_2)} \\[2mm]
& + \frac{\mu_{12}\mu_{23}\mu_{34}\,e^{(\mu_1-\mu_{46})f}(e^{-(\delta+\mu_1)f} - e^{-(\delta+\mu_1)n})}{(\mu_1 - \mu_2)(\mu_3 - \mu_{46})(\mu_1 - \mu_{46})(\delta + \mu_1)}.
\end{aligned}
$$

8

Indexing benefits in insurance packages

8.1 INTRODUCTION

This chapter deals with the problem of indexing benefits and premiums in insurance packages, which may in particular include disability covers. The actuarial approach we will present can also be used in analysing indexing problems in collective life insurance contracts.

From a financial point of view, we adopt a conventional approach, in the sense that fixed, non-stochastic interest rates are considered. A simple and flexible model allows us to take into account ex-post adjustments, based on several indexes (such as inflation rates, yields from investments, etc.). From a modelling point of view, the indexing problem is embedded in a Markov multiple-state framework. A time-continuous approach is adopted.

The chapter is organized as follows. Section 8.2 constitutes the theoretical body of the chapter. In this section we first introduce the concept of 'adjustment rates' and then we derive some 'mean formulae', in order to express the benefit adjustment rates as weighted arithmetic means of the adjustment rates affecting the premiums and the reserve. Section 8.3 describes some practical aspects of benefit indexing in insurance packages; in particular, the effect of constraints in benefit indexing is emphasized. Finally, section 8.4 is devoted to the illustration of indexing problems related to benefits commonly present in insurance packages and in life insurance collective contracts (as well as in pension schemes).

Since several types of benefits will be considered (life annuities, disability annuities, reversionary annuities, death lump sum benefits, etc.), in this chapter we will use the general notation defined in Chapter 1.

8.2 LINKING BENEFITS AND PREMIUMS TO SOME INDEX – A FORMAL STATEMENT

We assume the following types of benefits:

- a continuous annuity benefit during stays in state j; let the rate of annuity benefit at time u be $b_j(u)$ per time unit if $S(u) = j$;
- a lump sum at each transition $j \to k$; let the amount be $c_{jk}(u)$ if the transition occurs at time u.

As far as premiums are concerned, we consider two types of premiums:

- premiums paid on a continuous basis during stays in state i; let the rate of premium at time u be $p_i(u)$ per time unit if $S(u) = i$;
- a single premium at time 0, $\pi_1(0)$, or single recurrent premiums, $\pi_i(u)$, at integer times $u = 0, 1, 2, \ldots$, if the insured is then in state i.

In this section we assume, for the sake of simplicity (but without loss of generality), that:

- for each transition $j \to k$, $c_{jk}(u) = c_{jk} > 0$ for any u, or $c_{jk}(u) = 0$ for any u;
- for each state, j, $b_j(u) = b_j > 0$ for any u, or $b_j(u) = 0$ for any u;
- for each state, i, $p_i(u) = p_i > 0$ for any u, or $p_i(u) = 0$ for any u;

With these assumptions, the reserve $\bar{V}_i(t)$ (concerning the case of continuous premiums) can be expressed as follows:

$$\bar{V}_i(t) = \sum_j b_j \bar{a}_{ij}(t, n) + \sum_j \sum_{k:k \neq j} c_{jk} \bar{A}_{ijk}(t, n) - \sum_j p_j \bar{a}_{ij}(t, n) \qquad (8.1)$$

where sums are extended over the proper indexes and n denotes the term of the policy (possibly $n = +\infty$). Equation (8.1) constitutes a particular case of equation (1.132) (see section 1.11 in Chapter 1). More general situations shall be considered in the following sections.

Suppose that at a given integer time t ($t = 1, 2, \ldots, n - 1$) some adjustments of the benefits are made. Note that at time $t = n$ no adjustment is meaningful, since no maturity benefit has been planned. Assume that, for any state j and any transition $j \to k$:

$$b_j \;\Rightarrow\; b_j(1 + g^{(b_j)}) \qquad (8.2)$$

$$c_{jk} \;\Rightarrow\; c_{jk}(1 + g^{(c_{jk})}) \qquad (8.3)$$

where the gs denote the rates of adjustment. In order to preserve the equivalence, the premiums and (possibly) the reserve must be amended. Let $S(t) = i$ be the state occupied at time t. Let $g^{(p_h)}$ and $g^{(\bar{V}_i)}$ denote the rate of amendment of the continuous premiums and the reserve, respectively. Of course, the premium adjustment is meaningful if there are still premiums to pay in some state, i.e. if there is at least a function p_h which is positive. Then, the equivalence condition (in the case of continuous premiums) can be written as follows:

$$\sum_j b_j(1 + g^{(b_j)})\bar{a}_{ij}(t, n) + \sum_j \sum_{k:k \neq j} c_{jk}(1 + g^{(c_{jk})})\bar{A}_{ijk}(t, n)$$

$$= \bar{V}_i(t)(1 + g^{(\bar{V}_i)}) + \sum_h p_h(1 + g^{(p_h)})\bar{a}_{ih}(t, n). \qquad (8.4)$$

Since the equivalence principle must be fulfilled by the expected present values of benefits and premiums and by the reserve at time t before adjustments, the equivalence condition can be written as follows:

$$\sum_j b_j g^{(b_j)}\bar{a}_{ij}(t, n) + \sum_j \sum_{k:k \neq j} c_{jk} g^{(c_{jk})}\bar{A}_{ijk}(t, n)$$

$$= \bar{V}_i(t)g^{(\bar{V}_i)} + \sum_h p_h g^{(p_h)}\bar{a}_{ih}(t, n). \qquad (8.5)$$

For example: the adjustment rates $g^{(b_j)}$ and $g^{(c_{jk})}$ can express some linking of the benefits to a price index; the rate $g^{(\bar{V}_i)}$ can express the revaluation of the reserve supported by the insurer with the interest surplus; then, the rates $g^{(p_h)}$ represent the premium amendments consistent with the equivalence principle.

From the equivalence condition (8.5), dividing both sides by the quantity

$$\sum_j b_j \bar{a}_{ij}(t, n) + \sum_j \sum_{k:k \neq j} c_{jk}\bar{A}_{ijk}(t, n)$$

the following equation is derived:

$$\frac{\sum_j b_j g^{(b_j)}\bar{a}_{ij}(t, n) + \sum_j \sum_{k:k \neq j} c_{jk} g^{(c_{jk})}\bar{A}_{ijk}(t, n)}{\sum_j b_j \bar{a}_{ij}(t, n) + \sum_j \sum_{k:k \neq j} c_{jk}\bar{A}_{ijk}(t, n)}$$

$$= \frac{\bar{V}_i(t)g^{(\bar{V}_i)} + \sum_h p_h g^{(p_h)}\bar{a}_{ih}(t, n)}{\sum_j b_j \bar{a}_{ij}(t, n) + \sum_j \sum_{k:k \neq j} c_{jk}\bar{A}_{ijk}(t, n)}. \qquad (8.6)$$

The left-hand side of equation (8.6) represents a weighted arithmetic mean of the adjustment rates of the benefits. Let us denote by $\tilde{g}$ this mean. Using equation (8.1), we find:

$$\tilde{g} = \frac{\bar{V}_i(t)g^{(\bar{V}_i)} + \sum_h p_h g^{(p_h)}\bar{a}_{ih}(t, n)}{\bar{V}_i(t) + \sum_h p_h \bar{a}_{ih}(t, n)}; \qquad (8.7)$$

that is, the mean rate $\tilde{g}$ is a weighted arithmetic mean of the adjustment rates of the premiums and the reserve. It is worthwhile to stress that the weights depend on the duration t.

Remark 1

In practice, it is common to have a maximum cumulative percentage increase in benefits. Alternatively, there can be a future purchase option (the so-called 'guaranteed insurability'), in which some adverse selection may occur on choice of the benefits to increase.

Remark 2
A systematic approach for analysing several index models for life insurance policies and pension schemes has been proposed by Pentikäinen (1968). Some of the ideas proposed in that paper are developed in Pitacco (1995b), in which the indexing problem is embedded in a Markov multiple-state framework.

Now let us consider some particular cases and a generalization.

8.2.1 The same adjustment rate for any benefit

Assume for any state j and any transition $j \to k$:

$$g^{(b_j)} = g^{(c_{jk})} = g$$

i.e. the same adjustment rate for any benefit. Then, we have from equations (8.6) and (8.7):

$$g = \frac{\bar{V}_i(t)g^{(\bar{V}_i)} + \sum_h p_h g^{(p_h)}\bar{a}_{ih}(t,n)}{\bar{V}_i(t) + \sum_h p_h \bar{a}_{ih}(t,n)}; \tag{8.8}$$

thus, the benefit adjustment rate is a weighted arithmetic mean of the rates of amendment of the premiums and of the reserve.

8.2.2 Just one benefit adjustment rate positive

Another particular case may be of practical interest. Let us assume that just one of the benefit adjustment rates is positive, whilst the others are equal to zero. For example, let us suppose that

$$g^{(b_1)} = g^* > 0;$$

in this case, from equation (8.5) we get:

$$g^* = \frac{\bar{V}_i(t)g^{(\bar{V}_i)} + \sum_h p_h g^{(p_h)}\bar{a}_{ih}(t,n)}{b_1 \bar{a}_{i1}(t,n)}. \tag{8.9}$$

Note that equation (8.9) exhibits the relationship between the actuarial value $b_1\bar{a}_{i1}(t,n)$ of the benefit under adjustment and its adjustment rate g^*, for a given set of rates $g^{(\bar{V}_i)}$, $g^{(p_h)}$ (note, however, that the actuarial value $b_1\bar{a}_{i1}(t,n)$ also affects the reserve value $\bar{V}_i(t)$). Roughly speaking, we can say that a benefit which is small in terms of its actuarial value (compared with the remaining benefits) will be amended according to a large adjustment rate.

8.2.3 Reserve adjustment rate equal to zero

Let us consider the following benefit arrangement. Continuous benefits b_2 and b_3 concern state 2 and state 3 respectively. Continuous premiums are paid while the insured is in state 1. Let us assume that at time t the insured is in state 1. At that time, benefit b_2 is adjusted and the new benefit is $b_2(1 + g^{(b_2)})$. Let us assume that the benefit amendment is funded by an increase in the premium, the reserve remaining unchanged. Formally, we get from (8.5):

$$g^{(b_3)} = \frac{p_1 g^{(p_1)} \bar{a}_{11}(t, n) - b_2 g^{(b_2)} \bar{a}_{12}(t, n)}{b_3 \bar{a}_{13}(t, n)}. \tag{8.10}$$

Relation (8.10) shows that, for given values of $g^{(b_2)}$ and $g^{(p_1)}$, the rate $g^{(b_3)}$ might be negative: in this case, the adjustment of benefit b_2 is partially funded by a decrease in benefit b_3.

8.2.4 Recurrent adjustments

In insurance practice, usually the adjustment does not occur once in a while; on the contrary, recurrent (in particular annual) adjustments must be planned when facing the problem of finding protection against losses caused by inflation, or the problem of linking some benefits to the progression of salaries. To this purpose, the model previously described can be generalized as follows.

Let us denote by

$$b_j[0], \quad c_{jk}[0], \quad p_h[0]$$

the benefits and the premiums stated at time $t = 0$, and by

$$b_j[t], \quad c_{jk}[t], \quad p_h[t]$$

the benefits and the premiums determined immediately after the adjustment at time t, $t = 1, 2, \ldots, n - 1$. The benefit adjustment at time t can be described as follows:

$$b_j[t - 1] \;\Rightarrow\; b_j[t] = b_j[t - 1](1 + g^{(b_j)}(t))$$

$$c_{jk}[t - 1] \;\Rightarrow\; c_{jk}[t] = c_{jk}[t - 1](1 + g^{(c_{jk})}(t))$$

where $g^{(b_j)}(t)$ and $g^{(c_{jk})}(t)$ denote the adjustment rates at time t affecting benefits. The benefit adjustment must be faced by premium and reserve adjustments, which can be expressed as follows, for $t = 1, 2, \ldots, n - 1$:

$$p_h[t - 1] \;\Rightarrow\; p_h[t] = p_h[t - 1](1 + g^{(p_h)}(t))$$

$$\bar{V}_i(t) \;\Rightarrow\; \bar{V}_i(t^+) = \bar{V}_i(t)(1 + g^{(\bar{V}_i)}(t))$$

where $g^{(P_h)}(t)$ and $g^{(\bar{V}_i)}(t)$ denote the adjustment rates at time t affecting premiums and reserve respectively; the meaning of $\bar{V}_i(t^+)$ is obvious.

The adjustment rates must satisfy, at any time t, the equivalence condition (8.5). All the preceding equations (i.e. (8.6) to (8.10)) still hold. Of course, for some t all the adjustment rates might be equal to zero.

8.3 PRACTICAL ASPECTS OF BENEFIT INDEXING IN INSURANCE PACKAGES

The following benefits are commonly present in insurance packages, in collective life insurance contracts, in (possibly insured) occupational pension schemes, in deposit administration plans, etc.:

1. retirement benefits, in the form of single life annuities;
2. death benefits, consisting in a lump sum and/or in an annuity to be paid to the widow

Moreover, the following benefits can be included in insurance contracts and pension schemes (and in the last case they are often called 'ancillary benefits'):

3. disability benefits, in the form of disability annuities, covering a wide range of disablement possibilities, from short-term sickness to permanent disability (see section 3.1 in Chapter 3);
4. medical expense covers;
5. long-term care benefits, in the form of annuities paid to the elderly who need nursing and/or medical care (see Chapter 6);
6. critical illness covers, which provide the insured with a lump sum (or a sequence of lump sums) when he contracts an illness included in a set of dread diseases specified by the cover conditions (see Chapter 5).

As far as the funding of benefits in pension schemes is concerned, several practical possibilities can be considered. First, a distinction has to be made between:

1. funding methods based on an individual equivalence principle, according to which premiums must be determined for each insured in such a way as to meet the actuarial value of the benefits;
2. funding methods based on an aggregate equivalence principle, according to which premiums (usually expressed as percentages of salaries) must globally face the benefits pertaining to the group (so that a solidarity effect is usually introduced).

We will restrict our discussion to methods of type 1, which are compulsory in individual life insurance contracts and very common in collective contracts (while methods of type 2 are rather common in non-insured occupational pension schemes). In methods of type 1, the insured benefits can be funded by:

- level premiums;
- single recurrent premiums.

Single recurrent premiums are often used to meet the costs related to death benefits and to disability benefits, while life annuities can be funded by level premiums as well as by single recurrent premiums.

Finding protection against losses caused by inflation is one of the most important problems in the design of life and health insurance products. Furthermore, when collective contracts are concerned, linking benefits to salary progression is often required. It follows that an increase in the projected measure of the final salary implies an increase in the projected pension benefit. Moreover, it is rather common for a pension scheme to be amended in order to improve benefits, in particular in respect of service after the date of amendment.

Hence, life and health insurance contracts (and pension schemes) should allow for:

1. adjustments of life annuity benefits in order to keep these benefits in line with career progression, salary progression and pension scheme rules;
2. possibly, amendments in other benefits in line with salary progression or with a price index.

The adjustments above mentioned must be financed by increasing the premium level and/or the reserve value. While the increase in the premium level is usually charged to the insured, the increase in the reserve value is commonly provided by the insurer using the interest surplus.

Note that the adjustments under 1 must be often interpreted as a 'constraint' in defining the relationships between premiums and benefits. In this case, the amendments under 2 must be consistent with a reasonable increase in the premium level and with the increase in the reserve value financed by the insurer; sometimes, a decrease in the benefit level could be required in order to keep the premium increase at a reasonable level.

In section 8.2 we illustrated in general terms how multiple-state models allow for a straightforward formal definition of the relationships among the adjustment rates. In section 8.4, several examples illustrate this possibility referring to some particular benefit packages.

8.4 SOME EXAMPLES

8.4.1 Retirement annuity; continuous premium

Let us consider a two-state model, in which:

state 1 = 'alive'   state 2 = 'dead'.

Let m denote the retirement date. A continuous annuity benefit b_1 will be paid to the pensioner (i.e. for $u > m$). A continuous premium p_1 is paid to the insurer while $u \leq m$. The equivalence principle is fulfilled if and only if:

$$p_1 \int_0^m v^u P_{11}(0, u)\, \mathrm{d}u = b_1 \int_m^{+\infty} v^u P_{11}(0, u)\, \mathrm{d}u \tag{8.11}$$

or

$$p_1 \bar{a}_{11}(0, m) = b_1 v^m P_{11}(0, m) \bar{a}_{11}(m, +\infty). \tag{8.11'}$$

At time t (if $S(t) = 1$), the prospective reserve is given by:

$$\bar{V}_1(t) = b_1 v^{m-t} P_{11}(t, m) \bar{a}_{11}(m, +\infty) - p_1 \bar{a}_{11}(t, m) \quad \text{if } t < m \tag{8.12}$$

$$\bar{V}_1(t) = b_1 \bar{a}_{11}(t, +\infty) \qquad\qquad\qquad\qquad\quad \text{if } t \geq m. \tag{8.13}$$

Let us suppose that at time t, $t < m$, an increase in the benefit is settled. Assume that $b_1(1 + g^{(b_1)})$ is the increased benefit. The equivalence condition (see equation (8.5)), slightly modified to consider the deferment of the benefit) requires premium and reserve amendments. Thus:

$$b_1 g^{(b_1)} v^{m-t} P_{11}(t, m) \bar{a}_{11}(m, +\infty) = \bar{V}_1(t) g^{(\bar{V}_1)} + p_1 g^{(p_1)} \bar{a}_{11}(t, m). \tag{8.14}$$

We include a brief comment which will help with the interpretation of the results presented in the following examples. A given increase in the benefit cannot be funded only by the same proportionate increase in the premium. Thus, an increase in the reserve is required, in the same proportionate amount. Otherwise, if no increase in the reserve were allowed (for example, because no interest surplus is credited to the insurer), a higher increase in the premium level would be necessary. Of course, an infinite set of solutions is theoretically available; in practice, the choice would be restricted to some solutions acceptable to both the insurer and the employer.

To conclude the example, let us assume that a benefit increase takes place at time $t > m$. In this case, we simply have

$$b_1 g^{(b_1)} \bar{a}_{11}(t, +\infty) = \bar{V}_1(t) g^{(\bar{V}_1)} \tag{8.15}$$

thus, $g^{(b_1)} = g^{(\bar{V}_1)}$. In practice, such a benefit increase will only occur because of a surplus credited by the insurer to the reserve.

8.4.2 Non-retroactive amendment in retirement annuity; continuous premium

Let us suppose that the retirement annuity is increased to improve benefits in respect of service after the date of amendment; this will be referred to as 'non-retroactive amendment'. Let s denote a projected measure of final salary. For example, s may be the average of the earnings in the final three or five years of service. We assume that the benefit b_1 is a fraction ρ of s for each year of pensionable service. Thus, prior to the amendment, we have:

$$b_1 = \rho s m. \tag{8.16}$$

At time t, $t < m$, the benefit fraction is changed non-retroactively from ρ to ρ^*, $\rho^* > \rho$. Hence, the amended benefit, b_1^*, is given by:

$$b_1^* = \rho s t + \rho^* s(m - t); \tag{8.17}$$

it follows that the annuity adjustment rate is given by:

$$g^{(b_1)} = \frac{b_1^*}{b_1} - 1 = \frac{\rho t + \rho^*(m - t)}{\rho m} - 1 < \frac{\rho^*}{\rho} - 1. \tag{8.18}$$

To fund this amendment, let us first choose the following possibility ('total service spread' funding). Let p_1' denote the premium which funds the benefit b_1^* over the total service period; we have (see equation (8.11')):

$$p_1' \bar{a}_{11}(0, m) = b_1^* v^m P_{11}(0, m) \bar{a}_{11}(m, +\infty). \tag{8.19}$$

Assume that at time t the premium is changed from p_1 to p_1'. Hence, $g^{(p_1)}$ is defined by

$$p_1' = p_1(1 + g^{(p_1)}). \tag{8.20}$$

Comparing equation (8.19) with (8.14) we see that

$$p_1' = p_1(1 + g^{(b_1)}) \tag{8.20'}$$

and then $g^{(p_1)} = g^{(b_1)}$.

From (8.14), it finally follows that the reserve must be amended in the same proportionate amount, i.e. $g^{(\bar{V}_1)} = g^{(p_1)} = g^{(b_1)}$. Thus, the reserve increases because of the amendment, although the benefit rate change is non-retroactive; this fact, which may be difficult to explain to a client who is not an actuary, is simply caused by the choice of the premium increase which implies $g^{(p_1)} = g^{(b_1)}$.

More acceptable results, from the point of view of the policyholder, can be achieved by using an alternative method to fund the pension scheme amendment. Let p_1'' denote the amended premium; let us set:

$$p_1'' = p_1 \frac{\rho^*}{\rho} \tag{8.21}$$

whence

$$g^{(p_1)} = \frac{\rho^*}{\rho} - 1 > g^{(b_1)}. \tag{8.22}$$

A premium increase factor, ρ^*/ρ, equal to the benefit increase factor (albeit the benefit amendment only works in respect of future service) is likely to be appealing to the client. It must be stressed that p_1'' is the premium meeting (if applied at time 0) the annuity which would be applicable if the amended benefit rate were applied to all service.

As far as the reserve is concerned, since $g^{(p_1)} > g^{(b_1)}$ we get (from (8.14)):

$$g^{(\bar{V}_1)} < g^{(b_1)}. \tag{8.23}$$

8.4.3 Retirement annuity; single recurrent premiums

Whereas a level premium approach is consistent (in the framework of pension schemes) with funding methods such as the 'individual entry age normal method', single recurrent premiums reflect the principle of methods such as the 'accrued benefit method'. Let us suppose that single recurrent premiums $\pi_1(u)$ are paid for $u = 0, 1, \ldots, m - 1$, if the insured is in state 1. Each single premium funds the benefit accrued during the year $(u, u + 1)$ of pensionable service. The equivalence principle requires that:

$$\sum_{u=0}^{m-1} v^u P_{11}(0, u) \pi_1(u) = b_1 v^m P_{11}(0, m) \bar{a}_{11}(m, +\infty). \tag{8.24}$$

At time t (if $S(t) = 1$), $t < m$, the prospective reserve is given by:

$$\bar{V}_1(t) = b_1 v^{m-t} P_{11}(t, m) \bar{a}_{11}(m, +\infty) - \sum_{u=\lceil t \rceil}^{m-1} v^{u-t} P_{11}(t, u) \pi_1(u) \tag{8.25}$$

where $\lceil t \rceil$ denote the smallest integer larger than or equal to t. For $t \geq m$ see equation (8.13).

Let us suppose that at time t, $t < m$, an increase in the benefit is settled. The equivalence condition requires premium and reserve amendments. Thus:

$$b_1 g^{(b_1)} v^{m-t} P_{11}(t, m) \bar{a}_{11}(m, +\infty) = \bar{V}_1(t) g^{(\bar{V}_1)} + g^{(p_1)} \sum_{u=\lceil t \rceil}^{m-1} v^{u-t} P_{11}(t, u) \pi_1(u).$$

$$\tag{8.26}$$

The comments presented under section 8.4.1 can help in interpreting equation (8.26).

8.4.4 Retirement – last survivor annuity; continuous premium

In this case the retirement annuity paid to the pensioner, say (x), will be paid after his death to the widow, say (y), until her death. Formally, a four-state model can be used to represent the benefit structure (see Example 11 in Chapter 1 in which, however, a reversionary annuity was only considered):

state 1: (x) alive, (y) alive; state 2: (x) dead, (y) alive;

state 3: (x) alive, (y) dead; state 4: (x) dead, (y) dead.

In states 1 and 3, a benefit $b_1 = b_3 = b$ is paid by the insurer for $u > m$; in state 2, a benefit b_2 is paid for any u. A premium $p_1 = p_3 = p$ is paid to the insurer during the stay in state 1 or state 3 for $u \leq m$. The following relations are self-explanatory:

$$p(\bar{a}_{11}(0, m) + \bar{a}_{13}(0, m)) = bv^m[P_{11}(0, m)(\bar{a}_{11}(m, +\infty) + \bar{a}_{13}(m, +\infty))$$
$$+ P_{13}(0, m)\bar{a}_{33}(m, +\infty)] + b_2\bar{a}_{12}(0, +\infty) \quad (8.27)$$

$$\bar{V}_1(t) = bv^{m-t}[P_{11}(t, m)(\bar{a}_{11}(m, +\infty) + \bar{a}_{13}(m, +\infty)) + P_{13}(t, m)\bar{a}_{33}(m, +\infty)]$$
$$+ b_2\bar{a}_{12}(t, +\infty) - p(\bar{a}_{11}(t, m) + \bar{a}_{13}(t, m)) \quad \text{if } t < m \quad (8.28)$$

$$\bar{V}_1(t) = b(\bar{a}_{11}(t, +\infty) + \bar{a}_{13}(t, +\infty)) + b_2\bar{a}_{12}(t, +\infty) \quad \text{if } t \geq m \quad (8.29)$$

$$\bar{V}_2(t) = b_2\bar{a}_{22}(t, +\infty) \quad (8.30)$$

$$\bar{V}_3(t) = bv^{m-t}P_{33}(t, m)\bar{a}_{33}(m, +\infty) - p\bar{a}_{33}(t, m) \quad \text{if } t < m \quad (8.31)$$

$$\bar{V}_3(t) = b\bar{a}_{33}(t, +\infty) \quad \text{if } t \geq m. \quad (8.32)$$

Let us suppose that at time t, $t < m$, the benefit concerning states 1 and 3 is changed from b to $b(1 + g^{(b)})$. If $S(t) = 3$, then we simply have:

$$bg^{(b)}v^{m-t}P_{33}(t, m)\bar{a}_{33}(m, +\infty) = \bar{V}_3(t)g^{(\bar{V}_3)} + pg^{(p)}\bar{a}_{33}(t, m) \quad (8.33)$$

and the comments presented under section 8.4.1 are applicable. On the contrary, if $S(t) = 1$, we have:

$$bg^{(b)}v^{m-t}[P_{11}(t, m)(\bar{a}_{11}(m, +\infty) + \bar{a}_{13}(m, +\infty)) + P_{13}(t, m)\bar{a}_{33}(m, +\infty)]$$

$$+ b_2g^{(b_2)}\bar{a}_{12}(t, +\infty) = \bar{V}_1(t)g^{(\bar{V}_1)} + pg^{(p)}(\bar{a}_{11}(t, m) + \bar{a}_{13}(t, m)); \quad (8.34)$$

hence, the amendment rate $g^{(b_2)}$ is also involved. In this case, we can suppose that the amendment in benefit b is a 'constraint', while the amendment in benefit b_2 must be consistent with an acceptable increase in premium; in particular a negative value for $g^{(b_2)}$ might be required in order to keep the premium increase at a reasonable level (for a given value of $g^{(\bar{V}_1)}$).

8.4.5 Retirement annuity plus death lump sum benefit; continuous premium

Let us consider the four-state model used for section 8.4.4. A premium $p_1 = p_3 = p$ is paid to the insurer during the stay in state 1 or state 3 for $u \leq m$. Now, a retirement annuity is paid to the pensioner until his death; a whole life assurance provides the widow with a lump sum. Hence, in states 1 and 3, a benefit $b_1 = b_3 = b$ is paid by the insurer for $u > m$; a lump sum is paid upon transition $1 \rightarrow 2$, for any u. The following relations are self-explanatory:

$$p(\bar{a}_{11}(0,m) + \bar{a}_{13}(0,m)) = bv'''[P_{11}(0,m)(\bar{a}_{11}(m,+\infty) + \bar{a}_{13}(m,+\infty))$$
$$+ P_{13}(0,m)\bar{a}_{33}(m,+\infty)] + c_{12}\bar{A}_{112}(0,+\infty) \quad (8.35)$$

$$\bar{V}_1(t) = bv'''^{-t}[P_{11}(t,m)(\bar{a}_{11}(m,+\infty) + \bar{a}_{13}(m,+\infty)) + P_{13}(t,m)\bar{a}_{33}(m,+\infty)]$$
$$+ c_{12}\bar{A}_{112}(t,+\infty) - p(\bar{a}_{11}(t,m) + \bar{a}_{13}(t,m)) \quad \text{if } t < m \quad (8.36)$$

$$\bar{V}_1(t) = b(\bar{a}_{11}(t,+\infty) + \bar{a}_{13}(t,+\infty)) + c_{12}\bar{A}_{112}(t,+\infty) \quad \text{if } t \geq m \quad (8.37)$$

$$\bar{V}_3(t) = bv'''^{-t}P_{33}(t,m)\bar{a}_{33}(m,+\infty) - p\bar{a}_{33}(t,m) \quad \text{if } t < m \quad (8.38)$$

$$\bar{V}_3(t) = b\bar{a}_{33}(t,+\infty) \quad \text{if } t \geq m. \quad (8.39)$$

Let us suppose that at time t, $t < m$, the benefit concerning states 1 and 3 is changed from b to $b(1 + g^{(b)})$. We will only consider $S(t) = 1$ (as regards $S(t) = 3$, the reader can refer to sections 8.4.1 and 8.4.4. Hence, we have:

$$bg^{(b)}v'''^{-t}[P_{11}(t,m)(\bar{a}_{11}(m,+\infty) + \bar{a}_{13}(m,+\infty)) + P_{13}(t,m)\bar{a}_{33}(m,+\infty)]$$

$$+ c_{12}g^{(c_{12})}\bar{A}_{112}(t,+\infty) = \bar{V}_1(t)g^{(\bar{V}_1)} + pg^{(p)}(\bar{a}_{11}(t,m) + \bar{a}_{13}(t,m)); \quad (8.40)$$

thus, the amendment rate $g^{(c_{12})}$ is also involved. Also now, we can suppose that the amendment in benefit b is a 'constraint', while the amendment in benefit c_{12} must be consistent with an acceptable increase in premium; in particular a negative value for $g^{(c_{12})}$ might be required in order to keep the premium increase at a reasonable level (for a given value of $g^{(\bar{V}_1)}$).

8.4.6 Retirement annuity plus disability annuity; continuous premium

The well-known three-state model (see Example 9 in Chapter 1) can be used to represent the benefit structure:

state 1: active;   state 2: disabled;   state 3: dead.

A retirement annuity b is paid to the pensioner, i.e. for $u > m$. A disability annuity b_2 is paid while the insured is in state 2, for $u \leq m$. A level premium is paid to the insurer for $u \leq m$, while in state 1.

The structure of benefits and premiums is extremely simplified, and perhaps somewhat unrealistic. In particular, note that the retirement annuity and the disability annuity are independent of past service time. Nevertheless, the assumed structure might actually relate to superannuation schemes, providing low amount benefits in addition to benefits provided by public or private pension schemes. Anyway, the purpose of illustration is probably better served by this very simple example.

Let us define the following actuarial values:

$$\bar{a}_{1.}(t, u) = \bar{a}_{11}(t, u) + \bar{a}_{12}(t, u) \tag{8.41}$$

$$\bar{a}_{2.}(t, u) = \bar{a}_{21}(t, u) + \bar{a}_{22}(t, u). \tag{8.42}$$

For given benefits b and b_2, the following equations determine the premium level and the reserves:

$$p_1 \bar{a}_{11}(0, m) = b_2 \bar{a}_{12}(0, m) + bv^m [P_{11}(0, m)\bar{a}_{1.}(m, +\infty)$$
$$+ P_{12}(0, m)\bar{a}_{2.}(m, +\infty)] \tag{8.43}$$

$$\bar{V}_1(t) = b_2 \bar{a}_{12}(t, m) + bv^{m-t}[P_{11}(t, m)\bar{a}_{1.}(m, +\infty)$$
$$+ P_{12}(t, m)\bar{a}_{2.}(m, +\infty)] - p_1 \bar{a}_{11}(t, m) \qquad \text{if } t < m \tag{8.44}$$

$$\bar{V}_1(t) = b\bar{a}_{1.}(t, +\infty) \qquad\qquad\qquad\qquad\qquad \text{if } t \geq m \tag{8.45}$$

$$\bar{V}_2(t) = b_2 \bar{a}_{22}(t, m) + bv^{m-t}[P_{21}(t, m)\bar{a}_{1.}(m, +\infty)$$
$$+ P_{22}(t, m)\bar{a}_{2.}(m, +\infty) - p_1 \bar{a}_{21}(t, m); \qquad \text{if } t < m \tag{8.46}$$

$$\bar{V}_2(t) = b\bar{a}_{2.}(t, +\infty). \qquad\qquad\qquad\qquad\qquad \text{if } t \geq m. \tag{8.47}$$

It is self-evident that the opportunity of adjustments mainly regards the case $S(t) = 1$ for $t < m$, and hence the reserve $\bar{V}_1(t)$ (with $t < m$) is involved. The following equation expresses the equivalence principle:

$$b_2 g^{(b_2)} \bar{a}_{12}(t, m) + bg^{(b)} v^{m-t}[P_{11}(t, m)\bar{a}_{1.}(m, +\infty) + P_{12}(t, m)\bar{a}_{2.}(m, +\infty)]$$

$$= \bar{V}_1(t) g^{(\bar{V}_1)} + p_1 g^{(p_1)} \bar{a}_{11}(t, m). \tag{8.48}$$

Also in this case, we can assume the amendment in benefit b as a 'constraint', whereas the amendment in benefit b_2 must be simply consistent with an acceptable increase in premium, given $g^{(\bar{V}_1)}$; in particular, negative values for $g^{(b_2)}$ might be imposed. Nevertheless, an amendment can be reasonable also when $S(t) = 2$. The equivalence principle then requires (for $t < m$):

$$b_2 g^{(b_2)} \bar{a}_{22}(t, m) + bg^{(b)} v^{m-t}[P_{21}(t, m)\bar{a}_{1.}(m, +\infty) + P_{22}(t, m)\bar{a}_{2.}(m, +\infty)]$$

$$= \bar{V}_2(t) g^{(\bar{V}_2)} + p_1 g^{(p_1)} \bar{a}_{21}(t, m). \tag{8.49}$$

In this case, it is reasonable to require $g^{(b_2)} \geq 0$ (as the disability annuity is in payment), reducing $g^{(b)}$ if necessary, consistently with an acceptable level of $g^{(p_1)}$, given $g^{(\bar{V}_2)}$.

8.5 SOME NUMERICAL EXAMPLES

In this section we present some numerical examples which illustrate the amendment mechanism, in the case of a single adjustment as well as in the case of recurrent (annual) adjustments.

The three-state model for permanent disability annuities (see Example 8 in Chapter 1) is used to represent the benefit structure:

state 1: active; state 2: disabled; state 3: dead.

The transition intensities are drawn from the Danish model (see section 3.3.1). The force of interest is $\delta = \ln 1.04$.

8.5.1 Permanent disability annuity plus death lump sum benefit; continuous premium

Let $x = 45$ be the age at entry. Let $n = 15$ be the policy term. Assume $b_2 = 1$, $c_{13} = c_{23} = c = 5$ (of course for $t \leq 15$). Assume that a continuous constant premium, p_1, is payable while the insured is in state 1, for 15 years at most. As far as the disability benefit is concerned, note that we have (according to the notation used in section 3.3.7) $m = n = r = 15$.

Formulae for premium and reserves are straightforward. We have:

$$p_1 \bar{a}_{11}(0, n) = b_2 \bar{a}_{12}(0, n) + c[\bar{A}_{113}(0, n) + \bar{A}_{123}(0, n)] \tag{8.50}$$

$$\bar{V}_1(t) = b_2 \bar{a}_{12}(t, n) + c[\bar{A}_{113}(t, n) + \bar{A}_{123}(t, n)] - p_1 \bar{a}_{11}(t, n) \tag{8.51}$$

$$\bar{V}_2(t) = b_2 \bar{a}_{22}(t, n) + c \bar{A}_{223}(t, n). \tag{8.52}$$

The actuarial value (at time 0) of the insured benefits is 0.7226. Then we find $p_1 = 0.0685$.

Let us consider possible adjustments at time $t = 5$, firstly assuming $S(5) = 1$. The relation among the rates of adjustment represents a particular case of equation (8.5). Hence we find:

$$b_2 g^{(b_2)} \bar{a}_{12}(t, n) + c g^{(c)}[\bar{A}_{113}(t, n) + \bar{A}_{123}(t, n)] = \bar{V}_1(t) g^{(\bar{V}_1)} + p_1 g^{(p_1)} \bar{a}_{11}(t, n). \tag{8.53}$$

In Table 8.1, for given values of $g^{(b_2)}$, $g^{(c)}$ and $g^{(p_1)}$, the resulting values of $g^{(\bar{V}_1)}$ are presented. It is interesting to note that high values of $g^{(\bar{V}_1)}$ are needed in order to meet the benefit adjustment when a low increase in premium is chosen. This is mainly due to the low amount of reserve generated by the disability cover and by the term assurance as well.

Table 8.1

$g^{(b_2)}$	$g^{(c)}$	$g^{(\bar{V}_1)}$	$g^{(p_1)}$
0.04	0.04	0.176	0.02
0.05	0.05	0.186	0.03
0.06	0.06	0.128	0.05
0.07	0.07	0.138	0.06
0.08	0.08	0.216	0.06

Table 8.2

$g^{(b_2)}$	$g^{(c)}$	$g^{(\bar{V}_1)}$	$g^{(p_1)}$
0.04	0.04	0.02	0.043
0.05	0.05	0.03	0.053
0.06	0.06	0.05	0.061
0.07	0.07	0.06	0.071
0.08	0.08	0.06	0.083

Table 8.3

$g^{(b_2)}$	$g^{(c)}$	$g^{(\bar{V}_1)}$	$g^{(p_1)}$
0.04	0.00	0.111	0.00
0.05	0.00	0.139	0.00
0.06	0.00	0.166	0.00
0.07	0.00	0.194	0.00
0.08	0.00	0.222	0.00

This fact also helps in interpreting Table 8.2, in which, for given values of $g^{(b_2)}$, $g^{(c)}$ and $g^{(\bar{V}_1)}$, the resulting values of $g^{(p_1)}$ are presented.

Moreover, note that $g^{(b_2)} = g^{(c)} = g$ and hence g is a weighted arithmetic mean of $g^{(\bar{V}_1)}$ and $g^{(p_1)}$ (see formula (8.8)).

Tables 8.3, 8.4 and 8.5 present some examples in which just one benefit adjustment rate (i.e. $g^{(b_2)}$) is positive.

In Table 8.6, from given values of $g^{(b_2)}$ and $g^{(p_1)}$ the adjustment rate $g^{(c)}$ is derived assuming $g^{(\bar{V}_1)} = 0$. Note that in all cases, although $g^{(p_1)} < g^{(b_2)}$, the increase in the premium level allows for an increase in the death benefit.

Now, assume $S(5) = 2$. Since only the permanent disability is considered, state 2 is strictly transient and hence premiums are no longer paid. Then, the benefit amendment must be completely faced by the increase in the reserve. Consider the results presented in Table 8.7. Just one benefit adjustment rate (i.e. $g^{(b_2)}$) is assumed to be positive; hence

Table 8.4

$g^{(b_2)}$	$g^{(c)}$	$g^{(\bar{V}_1)}$	$g^{(p_1)}$
0.04	0.00	0.00	0.016
0.05	0.00	0.00	0.020
0.06	0.00	0.00	0.024
0.07	0.00	0.00	0.028
0.08	0.00	0.00	0.033

Table 8.5

$g^{(b_2)}$	$g^{(c)}$	$g^{(\bar{V}_1)}$	$g^{(p_1)}$
0.04	0.00	0.014	0.014
0.05	0.00	0.018	0.018
0.06	0.00	0.022	0.022
0.07	0.00	0.025	0.025
0.08	0.00	0.028	0.028

Table 8.6

$g^{(b_2)}$	$g^{(c)}$	$g^{(\bar{V}_1)}$	$g^{(p_1)}$
0.04	0.005	0.00	0.02
0.05	0.013	0.00	0.03
0.06	0.035	0.00	0.05
0.07	0.043	0.00	0.06
0.08	0.037	0.00	0.06

the numerical results can be interpreted using formula (8.9), which in this particular cases reduces to:

$$g^{(b_2)} = \frac{\bar{V}_2(t)g^{(\bar{V}_2)}}{b_2\bar{a}_{22}(t, n)}. \tag{8.54}$$

From equation (8.52) we obtain $\bar{V}_2(t) > b_2\bar{a}_{22}(t, n)$ and then $g^{(b_2)} > g^{(\bar{V}_2)}$.

Tables 8.8 to 8.11 relate to recurrent (annual) adjustments. Many comments concerning the previous examples also apply to the examples describing recurrent adjustments, and hence can be helpful in interpreting the numerical results.

It is worthwhile to stress that all the relations among the adjustment rates involve parameters (i.e. actuarial values) which depend on t – the duration since policy issue. As a result, for example in Table 8.8 we see that the rate $g^{(\bar{V}_1)}(t)$ depends on the duration, even though the rates

Table 8.7

$g^{(b_2)}$	$g^{(c)}$	$g^{(\bar{V}_2)}$
0.04	0.00	0.038
0.05	0.00	0.047
0.06	0.00	0.057
0.07	0.00	0.066
0.08	0.00	0.076

Table 8.8

t	$g^{(b_2)}$	$g^{(c)}$	$g^{(\bar{V}_1)}$	$g^{(p_1)}$
1	0.05	0.05	0.363	0.03
2	0.05	0.05	0.179	0.03
3	0.05	0.05	0.129	0.03
4	0.05	0.05	0.105	0.03
5	0.05	0.05	0.092	0.03
6	0.05	0.05	0.084	0.03
7	0.05	0.05	0.078	0.03
8	0.05	0.05	0.074	0.03
9	0.05	0.05	0.071	0.03
10	0.05	0.05	0.070	0.03
11	0.05	0.05	0.069	0.03
12	0.05	0.05	0.069	0.03
13	0.05	0.05	0.070	0.03
14	0.05	0.05	0.073	0.03

$g^{(b_2)}(t)$, $g^{(c)}(t)$ and $g^{(p_1)}(t)$ are assumed to be constant throughout the policy period.

Table 8.11 in particular shows that a decrease in the death benefit is required in order to fund the increase in the disability benefit. Of course, this is due to the fact that $g^{(\bar{V}_1)}(t) = g^{(b_2)}(t)$ and $g^{(p_1)} = 0$ for any t.

Table 8.9

t	$g^{(b_2)}$	$g^{(c)}$	$g^{(\bar{V}_1)}$	$g^{(p_1)}$
1	0.04	0.04	0.03	0.0403
2	0.04	0.04	0.03	0.0406
3	0.04	0.04	0.03	0.0409
4	0.04	0.04	0.03	0.0412
5	0.04	0.04	0.03	0.0414
6	0.04	0.04	0.03	0.0417
7	0.04	0.04	0.03	0.0419
8	0.04	0.04	0.03	0.0420
9	0.04	0.04	0.03	0.0421
10	0.04	0.04	0.03	0.0421
11	0.04	0.04	0.03	0.0421
12	0.04	0.04	0.03	0.0419
13	0.04	0.04	0.03	0.0416
14	0.04	0.04	0.03	0.0412

Table 8.10

t	$g^{(b_2)}$	$g^{(c)}$	$g^{(\bar{V}_1)}$	$g^{(p_1)}$
1	0.04	0.0380	0.00	0.04
2	0.04	0.0363	0.00	0.04
3	0.04	0.0348	0.00	0.04
4	0.04	0.0336	0.00	0.04
5	0.04	0.0326	0.00	0.04
6	0.04	0.0319	0.00	0.04
7	0.04	0.0314	0.00	0.04
8	0.04	0.0312	0.00	0.04
9	0.04	0.0313	0.00	0.04
10	0.04	0.0318	0.00	0.04
11	0.04	0.0326	0.00	0.04
12	0.04	0.0338	0.00	0.04
13	0.04	0.0354	0.00	0.04
14	0.04	0.0374	0.00	0.04

8.5.2 Retirement annuity plus permanent disability annuity; continuous premium

Let us consider the benefit structure described in section 8.4.6. Let $x = 45$ be the age at entry and let $m = 15$. Assume that a constant continuous premium, p_1, is payable while the insured is in state 1, for 15 years at most.

Premium and reserves can be determined using formulae (8.41) to (8.47). The relations among the adjustment rates when $S(t) = 1$ or $S(t) = 2$ are respectively given by formulae (8.48) and (8.49). Note,

Table 8.11

t	$g^{(b_2)}$	$g^{(c)}$	$g^{(\bar{V}_1)}$	$g^{(p_1)}$
1	0.04	−0.024	0.04	0.00
2	0.04	−0.023	0.04	0.00
3	0.04	−0.022	0.04	0.00
4	0.04	−0.020	0.04	0.00
5	0.04	−0.019	0.04	0.00
6	0.04	−0.018	0.04	0.00
7	0.04	−0.017	0.04	0.00
8	0.04	−0.015	0.04	0.00
9	0.04	−0.014	0.04	0.00
10	0.04	−0.012	0.04	0.00
11	0.04	−0.011	0.04	0.00
12	0.04	−0.010	0.04	0.00
13	0.04	−0.008	0.04	0.00
14	0.04	−0.007	0.04	0.00

Table 8.12

$g^{(b)}$	$g^{(b_2)}$	$g^{(\bar{V}_1)}$	$g^{(p_1)}$
0.04	0.04	0.070	0.02
0.05	0.05	0.080	0.03
0.06	0.06	0.075	0.05
0.07	0.07	0.085	0.06
0.08	0.08	0.110	0.06

Table 8.13

$g^{(b)}$	$g^{(b_2)}$	$g^{(\bar{V}_1)}$	$g^{(p_1)}$
0.04	0.04	0.02	0.053
0.05	0.05	0.03	0.063
0.06	0.06	0.05	0.067
0.07	0.07	0.06	0.077
0.08	0.08	0.06	0.093

however, that now we consider only permanent disability. Hence all probabilities and actuarial values involving the transition $2 \rightarrow 1$ are equal to 0. So, for example, we have (see formulae (8.42) and (8.46) respectively):

$$\bar{a}_{2\cdot}(t,u) = \bar{a}_{22}(t,u) \qquad (8.55)$$

$$\bar{V}_2(t) = b_2\bar{a}_{22}(t,m) + bv^{m-t}P_{22}(t,m)\bar{a}_{22}(m,+\infty) \quad \text{if } t < m. \qquad (8.56)$$

The actuarial value (at time 0) of the insured benefits is 7.2346. Then we find $p_1 = 0.6854$.

Let us consider possible adjustments at time $t = 5$, assuming $S(5) = 1$. As stressed in section 8.4.6, it is self-evident that the opportunity of adjustments mainly concerns the case $S(t) = 1$, for $t < m$. Tables 8.12 to 8.17 illustrate several arrangements of adjustment rates, which can be easily interpreted also referring to comments in section 8.4.1. The results in

Table 8.14

$g^{(b)}$	$g^{(b_2)}$	$g^{(\bar{V}_1)}$	$g^{(p_1)}$
0.04	0.00	0.097	0.00
0.05	0.00	0.121	0.00
0.06	0.00	0.145	0.00
0.07	0.00	0.170	0.00
0.08	0.00	0.194	0.00

Table 8.15

$g^{(b)}$	$g^{(b_2)}$	$g^{(\bar{V}_1)}$	$g^{(P_1)}$
0.04	0.00	0.00	0.065
0.05	0.00	0.00	0.082
0.06	0.00	0.00	0.098
0.07	0.00	0.00	0.114
0.08	0.00	0.00	0.131

Table 8.16

$g^{(b)}$	$g^{(b_2)}$	$g^{(\bar{V}_1)}$	$g^{(P_1)}$
0.04	0.00	0.039	0.039
0.05	0.00	0.049	0.049
0.06	0.00	0.059	0.059
0.07	0.00	0.068	0.068
0.08	0.00	0.078	0.078

Table 8.17 illustrate how an increase in benefit b can be financed also by a decrease in benefit b_2.

Tables 8.18 and 8.19 relate to annual adjustments. Also in this case we restrict our attention to the case $S(t) = 1$, with $t < 15$. Many comments concerning the previous examples (see section 8.4.1) can be helpful in interpreting the numerical results.

As already pointed out, all the relations among the adjustment rates involve parameters (i.e. actuarial values) which depend on the duration since policy issue. As a result, for example, in Table 8.18 we see that the rate $g^{(\bar{V}_1)}(t)$ depends on the duration, although the rates $g^{(b)}(t)$, $g^{(b_2)}(t)$ and $g^{(P_1)}(t)$ are constant throughout the deferment period.

8.6 REFERENCES AND SUGGESTIONS FOR FURTHER READING

Although the problem of finding protection against losses caused by inflation and hence the problem of indexing benefits are of dramatic

Table 8.17

$g^{(b)}$	$g^{(b_2)}$	$g^{(\bar{V}_1)}$	$g^{(P_1)}$
0.04	−0.126	0.03	0.04
0.05	−0.282	0.03	0.05
0.06	−0.272	0.04	0.06
0.07	−0.427	0.04	0.07
0.08	−0.583	0.04	0.08

Table 8.18

t	$g^{(b)}$	$g^{(b_2)}$	$g^{(\bar{V}_1)}$	$g^{(p_1)}$
1	0.04	0.04	0.145	0.03
2	0.04	0.04	0.086	0.03
3	0.04	0.04	0.068	0.03
4	0.04	0.04	0.059	0.03
5	0.04	0.04	0.054	0.03
6	0.04	0.04	0.050	0.03
7	0.04	0.04	0.048	0.03
8	0.04	0.04	0.046	0.03
9	0.04	0.04	0.044	0.03
10	0.04	0.04	0.043	0.03
11	0.04	0.04	0.042	0.03
12	0.04	0.04	0.042	0.03
13	0.04	0.04	0.041	0.03
14	0.04	0.04	0.0405	0.03

importance in the design of life and health insurance products, the relevant actuarial literature is not very extensive.

The first paper dealing with a systematic approach for analysing and comparing several index models for life insurance policies and pension schemes seems to be due to Pentikäinen (1968).

An analysis of health insurance costs related to price index linking is proposed by Pitacco (1989).

Table 8.19

t	$g^{(b)}$	$g^{(b_2)}$	$g^{(\bar{V}_1)}$	$g^{(p_1)}$
1	0.04	0.04	0.03	0.041
2	0.04	0.04	0.03	0.042
3	0.04	0.04	0.03	0.043
4	0.04	0.04	0.03	0.045
5	0.04	0.04	0.03	0.047
6	0.04	0.04	0.03	0.049
7	0.04	0.04	0.03	0.051
8	0.04	0.04	0.03	0.055
9	0.04	0.04	0.03	0.059
10	0.04	0.04	0.03	0.065
11	0.04	0.04	0.03	0.073
12	0.04	0.04	0.03	0.087
13	0.04	0.04	0.03	0.113
14	0.04	0.04	0.03	0.188

Some of the ideas proposed by Pentikainen are developed in Pitacco (1995b), from which the present chapter is mainly drawn; embedding the indexing problem in a Markov multiple-state framework allows for a more general approach.

When collective contracts are concerned, linking benefits to salary progression is often required. The relevant actuarial problems are analysed by Linnemann (1994), from which the example in Section 8.4.2 has been drawn.

Bibliography

Abbreviations used

AB   *ASTIN Bulletin*
ARCH   *Actuarial Research Clearing House*
ASMDA   *Applied Stochastic Models and Data Analysis*
n ASTIN   *Proceedings of the nth ASTIN Colloquium*
BARAB   *Bulletin de l'Association Royale des Actuaires Belges*
BDGV   *Blätter der deutschen Gesellschaft für Versicherungsmathematik*
BTIAF   *Bulletin Trimestriel de l'Institute des Actuaires Français*
CMIRm   *Continuous Mortality Investigation Reports, No. m*
GIIA   *Giornale dell'Istituto Italiano degli Attuari*
IME   *Insurance: Mathematics & Economics*
JAP   *Journal of Actuarial Practice*
JIA   *Journal of the Institute of Actuaries*
JIASS   *Journal of the Institute of Actuaries Students' Society*
JSIAS   *Journal of Staple Inn Actuarial Society*
MVSV   *Mitteilungen der Vereinigung schweizerischer Versicherungsmathematiker*
NAAJ   *North American Actuarial Journal*
SA   *Skandinavisk Aktuarietidskrift/Scandinavian Actuarial Journal*
TIAANZ   *Transactions of the Institute of Actuaries of Australia and New Zealand*
TnICA   *Transactions of the nth International Congress of Actuaries*
TSA   *Transactions of the Society of Actuaries*
TSAR   *Transactions of the Society of Actuaries. Reports of Mortality, Morbidity and other Experience*

Alegre, A. (1990) *Valoracion actuarial de prestaciones relacionadas con la invalidez*, Publicacions Universitat de Barcelona, Barcelona.

Allerdissen, K., Drude, G. and Gebhardt, T. (1993) Zur Calculation von Dread Disease. *BDGV.*

Alting von Gesau, B.J.J. (1990) AIDS in the Eighties and Nineties. *Liber Amicorum, Kok-Van Klinken*, Institute of Actuarial Science and Econometrics, University of Amsterdam, 1–22.

Amsler, M.H. (1968) Les chaînes de Markov des assurances vie, invalidité et maladie. *T18ICA*, München, **5**, 731–746.

Amsler, M.H. (1988) Sur la modélisation de risques vie par les chaînes de Markov. *T23ICA*, Helsinki, **3**, 1–17.

Anderson, R.M., May, R.M., Medley, G.F. and Johnson, A.E. (1986) A preliminary study of the transmission dynamics of the Human Immunodeficiency Virus (HIV), the causative agent of AIDS. *I.M.A.J. Math. Appl. Med. & Biol.*, **3**, 229–263.

Anderson, W.J. (1991) *Continuous-time Markov chains. An applications-oriented approach*, Springer-Verlag, New York.

Bague, A. (1991) Private Long-Term Care Insurance. *Proceedings of 'Les enjeux de la prévoyance: un défi pour demain'*, Paris, **3**, 167–179.

Bailey, N.T.J (1975) *The mathematical theory of infectious diseases and its applications*, Griffin, London.

Barnhart, E.P. (1985) A new approach to premium, policy and claim reserves for health insurance. *TSA*, **37**, 13–41.

Bhattacharya, R.N. and Waymire, E.C. (1990) *Stochastic processes with applications*, Wiley, New York.

Beekman, J.A. (1990) An alternative premium calculation method for certain long-term care coverages. *ARCH*, 179–199.

Beekman, J.A. and Frye, W.B. (1991) Projections of active life expectancies. *ARCH*, 73–87.

Benjamin, S. (1992) Using the value of their houses to fund long term care for the elderly. *T24ICA*, Montreal, **4**, 19–34.

Benjamin, B. and Pollard, J.H. (1980) *The analysis of mortality and other actuarial statistics*, Heinemann, London.

Berger, A. (1925) *Die Prinzipien der Lebensversicherungstechnik*, Zweiter Teil, Springer, Berlin.

Bernoulli, D. (1766) Essai d'une nouvelle analyse de la mortalité causée par la petite vérole, et des avantages de l'inoculation pour la prévenir. *Hist. Acad. Roy. Sci.*, Année MDCCLX, Mémoires, 1–45.

Bjoraa, S.J. (1951) On a simplified method of investigation into disability insurance. *T13ICA*, Scheveningen, **3**, 216–223.

Bjoraa, S.J. (1960) Disability insurance in Norway. Recent experience. *T16ICA*, Bruxelles, **1**, 190–196.

Black, K. and Skipper, H.D. (1987) *Life Insurance*, Prentice Hall, Englewood Cliffs, New Jersey, USA.

Bond, D.J. (1963) Permanent sickness insurance. *JIASS*, **17**, 195–224.

Booth, P.M. (1996) *Long-term care for the elderly: a review of policy options.* Actuarial Research Paper No. 86, Department of Actuarial Science & Statistics, City University, London.

Bowers, N.L., Gerber, H.U., Hickman, J.C., Jones, D.A. and Nesbitt, C.J. (1986) *Actuarial mathematics*, The Society of Actuaries, Itasca, Illinois.

Brodt, H.R., Helm, E.B., Werner, A., Joetten, A., Bergmann, L., Kluwer, A. and Stille, W. (1986). Spontanverlauf der LAV/HTLV-III Infektion; Verlaufsbeobachtungen bei Personen aus AIDS-Risikogruppen. *Deutsche Medizinische Wochenschrift*, Stuttgart, **III**, 1175–1180.

Bull, O. (1980) Premium calculations for two disability lump sum contracts. *T21ICA*, Zürich-Lausanne, **3**, 45–51.

Chadburn, R.G., Cooper, D.R. and Haberman, S. (1995) *Actuarial Mathematics*, Institute and Faculty of Actuaries, Oxford.

Chuard, M. (1995) Emploi du taux instantané de reactivité. *T25ICA*, Brussels, **1**, 95–112.

Chuard, M. and Chuard, P. (1992) La réactivité des invalides dans les rentes futures d'invalidité. *MVSV*, 61–82.

Chuard, P. (1993) Modèles mathématiques pour actifs et invalides. *MVSV*, 251–268.

Chuard, P. (1995) Passage des probabilités aux taux instantanés. *T25ICA*, Brussels, **1**, 113–128.

Clark, G. and Baker, B. (1992) *Group Permanent Health Insurance*, Paper presented to the Staple Inn Actuarial Society, London.

Clark, G. and Dullaway, D. (1995) *PHI pricing*, Health and Care PHI Meeting, Institute of Actuaries, London.

Consael, R. and Sonnenschein, J. (1978) Théorie mathématique des assurances de personnes. *MVSV*, 75–93.

CMIR7 (1984) Sickness Experience 1975–78 for Individual PHI Policies. *Continuous Mortality Investigation Bureau*, The Institute of Actuaries and the Faculty of Actuaries.

CMIR8 (1986) Sickness Experience 1975–78 for Group PHI Policies. *Continuous Mortality Investigation Bureau*, The Institute of Actuaries and the Faculty of Actuaries.

CMIR12 (1991) The Analysis of Permanent Health Insurance Data. *Continuous Mortality Investigation Bureau*, The Institute of Actuaries and the Faculty of Actuaries.

Copas, J.B. and Haberman, S. (1983) Non parametric graduation using kernel methods. *JIA*, **110**, 135–156.

Cordeiro, I.M.F. (1995) Sensitivity analysis in a multiple state model for permanent health insurance. *CMIR14*, The Institute of Actuaries and the Faculty of Actuaries, Oxford, 129–153.

Corliss, G., Ball, L., Newton, M. and Van Slyke, G. (1996) Report of the Inter-company Subcommittee of the Long-Term-Care Experience Committee, Long-Term-Care intercompany study: 1984–1991 experience, 1993–94 *TSAR*, 43–121.

Courant, S.T. (1984) Design of Disability Benefits. *T22ICA*, Sydney, **2**, 391–400.

Cournot, A.A. (1843) *Exposition de la théorie des chances et des probabilités*, Librairie de L'Hachette, Paris.

Cowley, A. (1992) *Long-term care insurance, international perspective and actuarial considerations*. Publications of the Cologne Re., No. 22, Cologne.

Cowell, M.H. and Hoskins, W.H. (1988) AIDS, HIV mortality and life insurance. In *The impact of AIDS on life and health insurance companies. A guide for practising actuaries*. Report of the Society of Actuaries Task Force on AIDS. The Society of Actuaries, Itasca, Illinois.

Cox, D.R. and Miller, H.D. (1965) *The theory of stochastic processes*, Chapman & Hall, London.

Cummins, J.D., Smith, B.E., Vance, R.N. and Van Der Hei, H.L. (1982) *Risk classification in life insurance*, Kluwer, Boston.

Daboni, L. (1964) Modelli di rendite aleatorie. *GIIA*, **27**, 273–296.

Dash, A. and Grimshaw, D. (1990) *Dread Disease Cover. An Actuarial Perspective.* Paper presented to the Staple Inn Actuarial Society, London.

Daw, R.H. (1979) Smallpox and the double decrement table: a piece of actuarial pre-history. *JIA*, **106**, 229–318.

Daw, R.H. (1980) Johann Heinrich Lambert (1728–1777). *JIA*, **107**, 345–363.

Daykin, C.D. (1990) Epidemiology of HIV infection and AIDS. *JIA*, **117**, 51–94.

Daykin, C.D., Clark, P.N.S., Eves, M.J. *et al.* (1988a) AIDS Bulletin No. 3. Report from the Institute of Actuaries AIDS Working Party, Institute of Actuaries, London.

Daykin, C.D., Clark, P.N.S., Eves, M.J. *et al.* (1988b) The impact of HIV infection and AIDS on insurance in the United Kingdom. *JIA*, **115**, 727–837.

Daykin, C.D., Clark, P.N.S., Eves, M.J. *et al.* (1989) AIDS Bulletin No. 4. Report from the Institute of Actuaries AIDS Working Party, Institute of Actuaries, London.

Daykin, C.D., Clark, P.N.S., Eves, M.J. *et al.* (1990) Projecting the spread of AIDS in the United Kingdom: a sensitivity analysis. *JIA*, **117**, 95–133.

De Pril, N. (1989) The distributions of actuarial functions, *MVSV*, 173–184.

De Vylder, F. (1973) *Théorie générale des opérations d'assurances individuelles de capitalisation*, Offices des Assureurs de Belgique, Bruxelles.

Depoid, P. (1967) *Application de la statistique aux assurances accidents et dommages*, Berger-Levrault, Paris.

Dhaene, J. (1990) Distributions in life insurance. *AB*, **20**, 81–92.

Dillner, C.G. (1969) New bases for non-cancellable sickness insurance in Sweden. *SA*, **52**, 113–124.

Dillner, C.G. (1974) New bases for long term sickness insurance in Sweden from 1973. *SA*, 167–173.

Du Pasquier, L.G. (1912) Mathematische Theorie der Invaliditätversicherung. *MVSV*, **7**, 1–7.

Du Pasquier, L.G. (1913) Mathematische Theorie der Invaliditätversicherung. *MVSV*, **8**, 1–153.

Eagles, L.M. (1992) Some problems in pricing long term care contracts. *T24ICA*, Montreal, **4**, 421–427.

Ekhult, H. (1980) Technique and experience in sickness and disability insurance in Sweden. *T21ICA*, Zürich-Lausanne, **3**, 67–82.

Elandt-Johnson, R.C. and Johnson, N.L. (1980) *Survival models and data analysis*, J. Wiley and Sons, New York.

Fabrizio, E. and Grattan, W.K. (1994) *Pricing Dread Disease Insurance*, Institute of Actuaries of Australia.

Fennell, C. (1991) Long term care: the challenge for the insurance industry. *Proceedings of 'Les enjeux de la prévoyance: un défi pour demain'*, Paris, **3**, 189–202.

Finlaison, J. (1829) *Life annuities – Report of John Finlaison, Actuary of the National debt on the evidence and elementary facts on which the tables of life annuities were founded*, House of Commons, London.

Fix, E. and Neyman, J. (1951) A Simple Stochastic Model of Recovery, Relapse, Death and Loss of Patients. *Human Biology*, **23**, 205–241.

Forfar, D.O., McCutcheon, J.J. and Wilkie, A.D. (1988) On graduation by mathematical formulae. *JIA*, **115**, 1–149.

Francis, B., Green, M. and Payne, C. (1993) Eds. *The GLIM System Release 4 Manual*. Clarendon Press, Oxford.

Franckx, E. (1963) Essai d'une théorie opérationelle des risques markoviens. *Quaderni dell'Istituto di Matematica Finanziaria dell'Università di Trieste*, No. 11, Trieste.

Frankona Rückversicherungs-AG (1989) *Calculating Premiums for a Critical Illness Cover*, Munich.

Gail, M.H. and Brookmeyer, R. (1988) Methods for projecting course of acquired immunodeficiency syndrome epidemic. *Journal of National Cancer Institute*, **80**, 900–911.

Galbrun, H. (1933) *Théorie mathématique de l'assurance invalidité et de l'assurance nuptialité*, Traité du calcul des probabilités et de ses applications, **III**, Gauthier-Villars, Paris.

Gatenby, P. (1991) *Long Term Care*. Paper presented to the Staple Inn Actuarial Society, London.

Gatenby, P. and Ward, N. (1994) *Multiple State Modelling*. Paper presented to the Staple Inn Actuarial Society, London.

Gavin, J., Haberman, S. and Verrall, R. (1993) Moving weighted average graduation using kernel estimation. *IME*, **12**, 113–126.

Gavin, J., Haberman, S. and Verrall, R. (1994) On the choice of bandwidth for kernel graduation. *JIA*, **121**, 119–134.

Gavin, J., Haberman, S. and Verrall, R. (1995) Variable kernel graduation with a boundary correction. *TSA*, **47**, to appear.

Gerber, H.U. (1979) *An introduction to mathematical risk theory*, Huebner Foundation, Philadelphia.

Goford, J. (1985) The control cycle. Financial control of a life assurance company. *JIASS*, **28**, 99–114.

Gregorius, F.K. (1993) Disability insurance in the Netherlands. *IME*, **13**, 101–116.

Haberman, S. (1983) Decrement tables and the measurement of morbidity: I. *JIA*, **110**, 361–381.

Haberman, S. (1984) Decrement tables and the measurement of morbidity: II. *JIA*, **111**, 73–86.

Haberman, S. (1987) Long-term sickness and invalidity benefits: forecasting and other actuarial problems. *JIA*, **114**, 467–533.

Haberman, S. (1988) The central sickness rate: a mathematical investigation. *T23ICA*, Helsinki, **3**, 83–99.

Haberman, S. (1992) HIV, AIDS and the approximate calculation of life insurance functions, annuities and net premiums. *JIA*, **119**, 345–364.

Haberman, S. (1993) *HIV, AIDS, Markov chains and PHI*, Actuarial Research Paper No. 52, Department of Actuarial Science & Statistics, City University, London.

Haberman, S. (1995) HIV, AIDS, Markov processes and health and disability insurance. *JAP*, **3**, 51–75.

Hamilton-Jones, J. (1972) Actuarial aspects of long-term sickness insurance. *JIA*, **98**, 17–67.

Hamza, E. (1900) Note sur la théorie mathématique de l'assurance contre le risque d'invalidité d'origine morbide, sénile ou accidentelle. *T3ICA*, Paris, 154–203.

Hännikäinen, M. (1988) *The disability risk and underlying factors*. Studies of the Central Pension Security Institute, Helsinki.

Hare, D.J.P. and McCutcheon, J.J. (1991) *An introduction to profit testing*. Actuarial Education Service Special Note, Institute of Actuaries, London.

Hauser, P.L. and Litow, M.E. (1996) Long-term care market receives help. *Disability Newsletter*, **71**, 4.

Heistermann, B., Price, R. and Trunk, S. (1990) *Long-term care. Characteristics, opportunities and prospects*. Publications of the Cologne Re., No. 16, Cologne.

Hertzman, E. (1993) Summary of talk on 'Work of the CMI PHI Sub-Committee'. *IME*, **13**, 117–122.

Hesselager, O. (1994) A Markov model for loss reserving. *AB*, **24**, 183–193.

Hesselager, O. and Norberg, R. (1996) On probability distributions of present values in life insurance. *IME*, **18**, 35–42.

Hoem, J.M. (1969a) Markov chain models in life insurance. *BDGV*, **9**, 91–107.

Hoem, J.M. (1969b) Some notes on the qualifying period in disability insurance. *MVSV*, 301–317.

Hoem, J.M. (1972) Inhomogeneous semi-Markov processes, select actuarial tables, and duration-dependence in demography. Greville, T.N.E. (ed.), *Population Dynamics*, Academic Press, London, 251–296.

Hoem, J.M. (1977) A Markov chain model of working life tables, *SA*, 1–20.

Hoem, J.M. (1988) The versatility of the Markov chain as a tool in the mathematics of life insurance. *T23ICA*, Helsinki, **R**, 171–202.

Hoem, J.M. and Funck Jensen, U. (1982) Multistate life table methodology: a probabilist critique. Land, K.C. and Rogers, A. (eds), *Multidimensional Mathematical Demography*, Academic Press, London, 155–264.

Hoem, J.M., Riis, J. and Sand, R. (1971) Disability income benefits in group life insurance. *SA*, **54**, 190–203.

Hubbard, G.N. (1852) *De l'organisation des Sociétés de Prévoyance ou de Secours mutuels et des bases scientifiques sur lesquelles elles doivent être établies*, Guillaumin et Cie, Paris.

Isham, V. (1988) Mathematical modelling of the transmission dynamics of HIV, infection and AIDS: a review. *Journal of Royal Statistical Society Series A*, **151**, 5–30.

Jacob, M. (1934) Sui metodi di approssimazione per il calcolo dei premi nelle assicurazioni d'invalidità. *T10ICA*, Rome, **1**, 304–329.

Janssen, J. (1966) Application des processus semi-markoviens à un problème d'invalidité. *BARAB*, **63**, 35–52.

Jones, B.L. (1992) An analysis of long-term care data from Hamilton-Wentworth, Ontario. *ARCH*, 337–352.

Jones, B.L. (1993) Modelling multi-state processes using a Markov assumption. *ARCH* (Proceedings of the 27th Annual Research Conference, Iowa City, 1992), 239–248.

Jones, B.L. (1994) Actuarial calculations using a Markov model. *TSA*, **46**, 227–250.

Jones, B.L. (1995) A stochastic population model for high demand CCRCs. *IME*, **16**, 69–77.

Jones, B.L. (1996) Transient results for a high demand CCRC model. *SA*, 165–182.

Jones, B.L. (1997a) Stochastic models for continuing care retirement communities. *NAAJ*, **1**(1), 50–73.

Jones, B.L. (1997b) Methods for the analysis of CCRC data. *NAAJ*, **1**(2), 40–54.

Jones, B.L. and Willmot, G.E. (1993) An open group long-term care model. *SA*, 161–172.

Jordan, C.W. (1982) *Life Contingencies*, The Society of Actuaries, Chicago.

Källström, L. (1990) Long-term individual sickness insurance according to the 1984 Rules (G84). *Sverige Reinsurance Company – 75 Years*, Uppsala, 31–50.

Kaplan, E.L. and Meier, P. (1958) Nonparametric estimation for incomplete observations. *Journal of American Statistical Association*, **53**, 457–481.

Karup, J. (1893) *Die Finanzlage der Gothaischen Staatsdiener-Wittwen-Societät am 31 December 1980*, Heinrich Morchel, Dresden.

Kuikka, S., Lindqvist, C. and Voivalin, M. (1980) The framework of actuarial bases and its application in the Finnish employment pension scheme. *SA*, 1–23.

Lambert, J.H. (1772) *The mortality of smallpox in children* (Beyträge zum Gebrauche der Mathematik und deren Anwendung, Berlin, Vol. III, 568–599). Translation (1980) in: *JIA*, **107**, 351–363.

Lazarus, W. (1860) Letter. *JIA*, **8**, 351–356.

Levikson, B. and Mizrahi, G. (1994) Pricing long term care insurance contracts. *IME*, **14**, 1–18.

Linnemann, P. (1994) Bonus, salary increases and real value of pensions. *SA*, 99–118.

London, D. (1985) *Graduation: the revision of estimates*, Actex Publications, Winsted, CT.

London, D. (1988) *Survival models and their estimation*, Actex Publications, Winsted, CT.

Lörper, J., Lüttgen, D. and Trunk, S. (1991) Dread-Disease-Deckungen. Ein erweitertes Leistungsangebot in der Lebenversicherung – Rechnungsgrundlagen für Deutschland. *BDGV*, **20**, 169–182.

Lundberg, O. (1969) Methods of studying the risk process in disability insurance. *AB*, **5**, 267–273.

McCullagh, P. and Nelder, J.A. (1989) *Generalized linear models*, Chapman & Hall, London.

Mackay, G. (1993) Permanent health insurance. Overviews and market conditions in the UK. *IME*, **13**, 123–130.

Makeham, W.M. (1867) On the law of mortality. *JIA*, **13**, 325–358.

Makeham, W.M. (1875) On an application of the theory of the composition of decremental forces. *JIA*, **18**, 317–322.

Martin, J., Meltzer, H. and Elliot, D. (1988) *OPCS Surveys of disability in Great Britain: Report 1 – The prevalence of disability among adults.* OPCS HMSO, London.

Mattsson, P. (1930) Technical methods of insurance against sickness disability. *T9ICA*, Stockholm, **3**, 167–197.

Mattsson, P. (1956) New bases for non-cancellable sickness insurance in Sweden. *SA*, **39**, 198–215.

Mattsson, P. (1977) Some reflections on different disability models. *SA*, 110–118.

Mattsson, P. and Lundberg, O. (1957) Risk functions of Swedish sickness insurance. *T15ICA*, New York, **2**, 371–384.

Mattsson, P. and Unneryd, A. (1968) Experience bases and assessment of premiums in Swedish long term health insurance. *T18ICA*, München, **4**, 523–533.

Maynard, P. (1991) Long term care insurance: production for the future. *Proceedings of 'Les enjeux de la prévoyance: un défi pour demain'*, Paris, **3**, 231–241.

Medin, K. (1951) A function for smoothing tables of the duration of sickness. *SA*, **34**, 45–52.

Mercantile & General Reinsurance (1992) *Dread disease. The critical facts*, London.

Meredith, J. (1973) A markovian analysis of a geriatric ward. *Management Science*, **19**, 604–612.

Miller, J.H. (1976) *The underwriting and control of long-term disability insurance*, NRG Publications, Amsterdam.

Miller, J.H. (1984) Partial disability insurance. *T22ICA*, Sidney, **4**, 121–128.

Minor, E.H. (1968) Comments on U.S. sickness insurance and bases of premiums therefore. *T18ICA*, Munich, **4**, 561–578.

Münchener Rück (1989) *Dread disease insurance*, München.

Münchener Rück (1992) *Cover of the long term care risk*, München.

Münchener Rück (1993) *Problems of the disability risk*, München.

Neill, A. (1977) *Life contingencies*, Heinemann, London.

Nelson, W.A. (1972) Theory and applications of hazard plotting for censored failure data. *Technometrics*, **14**, 945–966.

Neyman, J. (1950) *A first course in probability and statistics*, Holt, New York.

Norberg, R. (1988) Select mortality: Possible explanations. *T23ICA*, Helsinki, **3**, 215–224.

Norberg, R. (1991) Reserves in life and pension insurance. *SA*, 3–24.

Norberg, R. (1995) A time-continuous Markov chain interest model with applications to insurance. *ASMDA*, **11**, 245–256.

Nuttall, S.R. (1992) Opportunities in long-term care insurance. *T24ICA*, Montreal, **4**, 239–254.

Nuttall, S.R. et al. (1994) Financing long-term care in Great Britain. *JIA*, **121**, 1–53.

O'Grady, F.T. (ed) (1988) *Individual Health Insurance*, Society of Actuaries, Schaumburg, Illinois, USA.

Olivieri, A. (1996) Sulle basi tecniche per coperture Long Term Care, *GIIA*, **49**, 87–116.

Ore, T.K., Sand, R. and Trier, G. (1964) New technical bases for life and pension insurance in Norway N1963, R1963, K1963. *SA*, **47**, 164–216.

Panjer, H.H. (1988) AIDS: Survival analysis of persons testing HIV+. *TSA*, **40**, 517–530.

Pentikäinen, T. (1968) Linking life and private pension insurance to the price index. *T18ICA*, Munich, **2**, 847–859.

Pitacco, E. (1989) Adjustment problems in permanent health insurance. *XXI ASTIN*, New York, 379–387.

Pitacco, E. (1994) LTC insurance. From the multistate model to practical implementations. *XXV ASTIN*, Cannes, 437–452.

Pitacco, E. (1995a) Actuarial models for pricing disability benefits: towards a unifying approach. *IME*, **16**, 39–62.

Pitacco, E. (1995b) Collective life insurance indexing. A multistate approach. *T25ICA*, Brussels, **1**, 335–356.

Pollard, J.H. (1995) *Long term care: demographic and insurance perspectives*, Macquarie University, School of Economic and Financial Studies, Research Paper No. 9/95.

PPP Lifetime Care (1996) *A world-wide review of long term care.*

Price, R. (1792) *Observations on reversionary payments on schemes for providing annuities for widows and for persons in old age and on the method of calculating the values of assurances on lives* (5th edition), T. Cadell, London.

Propp, J.S. (1992) Meeting the need for long-term care insurance in Israel. *T24ICA*, Montreal, **4**, 303–322.

Puzey, A.S. (1997) *A general theory of mortality rate estimators*, Actuarial Research Paper No. 98, Department of Actuarial Science & Statistics, City University, London.

Ramlau-Hansen, H. (1988) The emergence of profit in life insurance. *IME*, **7**, 225–236.

Ramlau-Hansen, H. (1991) Distribution of surplus in life insurance. *AB*, **21**, 57–71.

Ramsay, C.M. (1989) AIDS and the calculation of life insurance functions. *TSA*, **41**, 393–422.

Redfield, R., Wright, D.C. and Tramont, E.C. (1986) The Walter Reed staging classification for HTLV-III/LAV infection. *New England Journal of Medicine*, **314**(2), 131–132.

Renshaw, A.E. (1991) Actuarial graduation practice and generalized linear and non-linear models. *JIA*, **118**, 295–312.

Renshaw, A.E. (1992) Joint modelling for actuarial graduation and duplicate policies. *JIA*, **119**, 69–85.

Renshaw, A.E. and Haberman, S. (1993) Graduations associated with a multiple state model for permanent health insurance. *Proceedings European Summer School on Health Insurance*, Brussels.

Renshaw, A.E. and Haberman, S. (1995) On the graduation associated with a multiple state model for permanent health insurance. *IME*, **17**, 1–17.

Report of the Committee on Individual Health Insurance (1990) Experience under individual disability loss-of-time policies: 1984–85. 1985–86–87 *TSAR*, 71–195.

Report of the Long-Term-Care Experience Committee (1992) 1985 National Nursing Home Survey Utilization Data. 1988–89–90 *TSAR*, 101–164.

Rickayzen, B.D. (1997) *A sensitivity analysis of the parameters used in a PHI multiple state model*, Actuarial Research Paper 103, Department of Actuarial Science & Statistics, City University, London.

Ross, S.M. (1983) *Introduction to probability models*, Academic Press, New York.

Royston, P. and Altman, D.G. (1994) Regression using fractional polynomials of continuous covariates: parsimonious parametric modelling, *Applied Statistics*, **43**, 429–467.

Sand, R. (1968) Disability pension insurance. A new method for calculation of premiums. *T18ICA*, München, **1**, 529–538.

Sand, R. and Riis, J. (1980) Some theoretical aspects and experience with a simplified method of premium rating in disability insurance. *T21ICA*, Zürich-Lausanne, **3**, 251–263.

Sanders, A.J. and Silby, N.F. (1988) Actuarial aspects of PHI in the U.K. *JSIAS*, **31**, 1–57.

Sandström, A. (1990) Long-term health insurance. *Sverige Reinsurance Company – 75 Years*, Uppsala, 101–140.

Sansom, R.J. and Waters, H.R. (1988) Permanent health insurance in the U.K.: the mathematical model and the statistical analysis of the data. *T23ICA*, Helsinki, **3**, 323–339.

Seal, H.L. (1970) Probability distributions of aggregate sickness duration, *SA*, **53**, 193–204.

Seal, H.L. (1977) Studies in the history of probability and statistics. XXXV. Multiple decrements or competing risks. *Biometrika*, **64**, 429–439.

Segerer, G. (1993) The actuarial treatment of the disability risk in Germany, Austria and Switzerland. *IME*, **13**, 131–140.

Sharp, K.P. (1993) Funding methods and pension plan amendments. *JAP*, **1**, 111–126.

Simonsen, W. (1936) Über die Grundlagen der Invaliditätsversicherung, *SA*, **19**, 27–51.

Söderström, L.G. (1980) Actuarial methods and results in the sickness insurance AGS. *T21ICA*, Zürich-Lausanne, **3**, 293–305.

Sprague, T.B. (1879) On the construction of a combined marriage and mortality table from observations made as to the rates of marriage and mortality among any body of men, *JIA*, **31**, 406–446.

Steffensen, J.F. (1949) On technical functions of invalidity insurance. *SA*, **32**, 160–175.

Steffensen, J.F. (1950) More about invalidity functions. *SA*, **33**, 193–202.

Stoltz, G. (1930) Some actuarial aspects regarding sickness insurance. *T9ICA*, Stockholm, **3**, 198–214.

Sverdrup, E. (1965) Estimates and test procedures in connection with stochastic models for deaths, recoveries and transfers between different states of health. *SA*, **48**, 184–211.

Swiss Re Group (1982) *Disability Handbook*, Swiss Reinsurance Company, Zürich.

Timmer, H.G. *et al.* (1990) *Technische Methoden der privaten Kranken versicherung in Europa: Marktverhältnisse und Wesensmerkmale der Versicherungstechnik.* Schriftenreihe Angewandte Versicherungsmathematik, Heft 23. Deutsche Gesellschaft für Versicherungsmathematik, Karlsruhe.

Tolley, H.D. and Manton, K.G. (1983) Multiple cause models of disease dependency. *SA*, 211–226.

Tuomikoski, J. (1988) On modelling disability by mixed exponentials. *T23ICA*, Helsinki, **3**, 423–444.

Türler, H. (1970) Zur Theorie der Invaliditätsversicherung. *MVSV*, **70**, 285–295.

Turner, M. (1988) *PHI reserving.* Paper presented to the Staple Inn Actuarial Society, London.

Vaupel, J.W., Manton, K.G. and Stallard, E. (1979) The impact of heterogeneity in individual frailty on the dynamics of mortality. *Demography*, **16**, 439–454.

Walker, B. (1992) Geronth insurance and geriatric assessment. *T24ICA*, Montreal, **4**, 355–374.

Waters, H.R. (1984) An approach to the study of multiple state models. *JIA*, **111**, 363–374.

Waters, H.R. (1989) Some aspects of the modelling of permanent health insurance. *JIA*, **116**, 611–624.

Waters, H.R. (1992) Non-reported claims in long term sickness insurance. *T24ICA*, Montreal, **2**, 335–342.

Waters, H.R. and Wilkie, A.D. (1987) A short note on the construction of life tables and multiple decrement tables. *JIA*, **113**, 569–580.

Watson, A.W. (1903) *Sickness and mortality experience of the Independent Order of Oddfellows Manchester Unity Friendly Society during the five years 1893–1897*, I.O.O.F.M.U., Manchester.

Westwood, W.G. (1972) The actuarial aspects of disability income insurance. *TIAANZ*, 65–111.

Wiegand, A. (1859) *Mathematische Grundlagen für Eisenbahn-Pensionkassen*, H.W. Schmidt, Halle.

Wilkie, A.D. (1988a) An actuarial model for AIDS. *JIA*, **115**, 839–853.

Wilkie, A.D. (1988b) Markov models for combined marriage and mortality tables. *T23ICA*, Helsinki, **3**, 473–486.

Wilkie, A.D. (1989) Population projection for AIDS using an actuarial model. *Phil. Trans. Royal Soc. B*, **325**, 99–112.

Winklevoss, H.E. and Powell, A.V. (1984) *Continuing care retirement communities. An empirical, financial, and legal analysis*, R.D. Irwin, Homewood, Ill.

Wolthuis, H. (1994) *Life insurance mathematics (The Markovian model)*, CAIRE Education Series No. 2, Bruxelles.

Zwinggi, E. (1958) *Versicherungsmathematik*, Birkhauser, Basel.

Index